Introduction to Nuclear and Particle Physics

Solutions Manual for Second Edition

Introduction to Nuclear and Particle Physics

Solutions Manual for Second Edition

C Bromberg
Michigan State University, USA

A Das
University of Rochester, USA

T Ferbel
University of Rochester, USA

NEW JERSEY • LONDON • SINGAPORE • BEIJING • SHANGHAI • HONG KONG • TAIPEI • CHENNAI

Published by

World Scientific Publishing Co. Pte. Ltd.

5 Toh Tuck Link, Singapore 596224

USA office: 27 Warren Street, Suite 401-402, Hackensack, NJ 07601

UK office: 57 Shelton Street, Covent Garden, London WC2H 9HE

British Library Cataloguing-in-Publication Data
A catalogue record for this book is available from the British Library.

First published 2006
Reprinted 2008, 2012

INTRODUCTION TO NUCLEAR AND PARTICLE PHYSICS
Solutions Manual for Second Edition

ISBN-13 978-981-256-744-4 (pbk)
ISBN-10 981-256-744-5 (pbk)

Printed by FuIsland Offset Printing (S) Pte Ltd Singapore

Contents

Contents

1. Rutherford Scattering

Problem 1.1 *Using Eq.* (1.38) *calculate the approximate total cross sections for Rutherford scattering of a* 10 MeV α*-particle from a lead nucleus for impact parameters* b *less than* 10^{-12}, 10^{-10} *and* 10^{-8} cm. *How well do these agree with the values of* πb^2?

There are various ways of doing this problem. We will list below two very simple methods.

Method I. In general, the total cross section for Rutherford scattering is given by (see Eq. (1.38) in the text)

$$\sigma_{\mathrm{TOT}} = 8\pi \left(\frac{ZZ'e^2}{4E}\right)^2 \int_0^1 \frac{\mathrm{d}\left(\sin\frac{\theta}{2}\right)}{\left(\sin\frac{\theta}{2}\right)^3}. \tag{1.1}$$

However, if the impact parameter is restricted to a finite range, say $b \leq b_0$, then we can write the total cross section as

$$\sigma_{\mathrm{TOT}}(b_0) = 8\pi \left(\frac{ZZ'e^2}{4E}\right)^2 \int_{\theta_{b_0}}^1 \frac{\mathrm{d}\left(\sin\frac{\theta}{2}\right)}{\left(\sin\frac{\theta}{2}\right)^3}, \tag{1.2}$$

where θ_{b_0} is the scattering angle corresponding to the impact parameter b_0 and is given by (see Eq. (1.32) of the text)

$$b_0 = \frac{ZZ'e^2}{2E} \cot\frac{\theta_{b_0}}{2}. \tag{1.3}$$

Carrying out the integration in (1.2), we obtain

$$\begin{aligned}
\sigma_{\mathrm{TOT}}(b_0) &= 8\pi\left(\frac{ZZ'e^2}{4E}\right)^2\left(-\frac{1}{2}\right)\left(1-\operatorname{cosec}^2\frac{\theta_{b_0}}{2}\right) \\
&= 4\pi\left(\frac{ZZ'e^2}{4E}\right)^2\cot^2\frac{\theta_{b_0}}{2} \\
&= \pi\left(\frac{ZZ'e^2}{2E}\cot\frac{\theta_{b_0}}{2}\right)^2 = \pi b_0^2, \qquad (1.4)
\end{aligned}$$

where we have used the identification in (1.3). It follows, therefore, that

b_0 (cm)	$\sigma_{\mathrm{TOT}}(b_0) = \pi b_0^2$ (cm^2)
10^{-12}	3.2×10^{-24}
10^{-10}	3.2×10^{-20}
10^{-8}	3.2×10^{-16}

Method II. An alternative method to obtain the same result is to note that the total cross section for Rutherford scattering can be written as

$$\begin{aligned}
\sigma_{\mathrm{TOT}} &= 8\pi\left(\frac{ZZ'e^2}{4E}\right)^2\int_0^1\frac{\mathrm{d}(\sin\frac{\theta}{2})}{(\sin\frac{\theta}{2})^3} \\
&= 4\pi\left(\frac{ZZ'e^2}{4E}\right)^2\int_0^1 \mathrm{d}\theta\,\cot\frac{\theta}{2}\operatorname{cosec}^2\frac{\theta}{2}. \qquad (1.5)
\end{aligned}$$

This can be converted into an integral over the impact parameters using the defining relationship (see Eqs. (1.32) and (1.36))

$$b = \frac{ZZ'e^2}{2E}\cot\frac{\theta}{2}, \qquad \frac{\mathrm{d}b}{\mathrm{d}\theta} = -\frac{ZZ'e^2}{4E}\operatorname{cosec}^2\frac{\theta}{2}, \qquad (1.6)$$

so that we can write

$$\begin{aligned}
\sigma_{\mathrm{TOT}} &= 4\pi\left(\frac{ZZ'e^2}{4E}\right)^2\int_0^\infty \mathrm{d}b\left(\frac{ZZ'e^2}{4E}\right)^{-1} b\left(\frac{ZZ'e^2}{2E}\right)^{-1} \\
&= 2\pi\int_0^\infty \mathrm{d}b\, b. \qquad (1.7)
\end{aligned}$$

This is true in general and can also be deduced from the definition of the cross section in Eq. (1.33) or (1.34) of the text. If impact parameters are smaller than some fixed value, say b_0, then the total cross section takes the form

$$\sigma_{\mathrm{TOT}}(b_0) = 2\pi \int_0^{b_0} \mathrm{d}b\, b = \pi b_0^2, \tag{1.8}$$

which is the same result as derived earlier.

Problem 1.2 *Prove that Eq.* (1.55) *follows from the relations in Eqs.* (1.53) *and* (1.54).

This problem can be solved directly from the relationship between the scattering angles in the laboratory and the CM frames. From Eq. (1.53) of the text we have

$$\cos\theta_{\mathrm{Lab}} = \frac{\cos\theta_{\mathrm{CM}} + \zeta}{(1 + 2\zeta\cos\theta_{\mathrm{CM}} + \zeta^2)^{1/2}}. \tag{1.9}$$

Through direct differentiation, it follows that

$$\begin{aligned}\frac{\mathrm{d}\cos\theta_{\mathrm{Lab}}}{\mathrm{d}\cos\theta_{\mathrm{CM}}} &= \frac{1}{(1 + 2\zeta\cos\theta_{\mathrm{CM}} + \zeta^2)^{1/2}} - \frac{\zeta(\cos\theta_{\mathrm{CM}} + \zeta)}{(1 + 2\zeta\cos\theta_{\mathrm{CM}} + \zeta^2)^{3/2}} \\ &= \frac{1 + \zeta\cos\theta_{\mathrm{CM}}}{(1 + 2\zeta\cos\theta_{\mathrm{CM}} + \zeta^2)^{3/2}},\end{aligned} \tag{1.10}$$

which leads to

$$\frac{\mathrm{d}\cos\theta_{\mathrm{CM}}}{\mathrm{d}\cos\theta_{\mathrm{Lab}}} = \frac{(1 + 2\zeta\cos\theta_{\mathrm{CM}} + \zeta^2)^{3/2}}{1 + \zeta\cos\theta_{\mathrm{CM}}}. \tag{1.11}$$

Let us note that

$$(1 + 2\zeta\cos\theta_{\mathrm{CM}} + \zeta^2) = (1 + \zeta\cos\theta_{\mathrm{CM}})^2 + \zeta^2\sin^2\theta_{\mathrm{CM}} > 0. \tag{1.12}$$

Since the differential cross sections in the two frames must be positive, the Jacobian connecting the two must also be positive. From

Eq. (1.54) of the text, this leads to the relationship (using the absolute value of the Jacobian)

$$\frac{d\sigma}{d\Omega_{\rm Lab}}(\theta_{\rm Lab}) = \frac{d\sigma}{d\Omega_{\rm CM}}(\theta_{\rm CM}) \left|\frac{d\cos\theta_{\rm CM}}{d\cos\theta_{\rm Lab}}\right|$$
$$= \frac{d\sigma}{d\Omega_{\rm CM}}(\theta_{\rm CM}) \frac{(1+2\zeta\cos\theta_{\rm CM}+\zeta^2)^{3/2}}{|1+\zeta\cos\theta_{\rm CM}|}, \quad (1.13)$$

which is the desired result (see Eq. (1.55) of the text).

Problem 1.3 *Sketch* $\cos\theta_{\rm Lab}$ *as a function of* $\cos\theta_{\rm CM}$ *for the nonrelativistic elastic scattering of particles of unequal mass, for the cases when* $\zeta = 0.05$ *and* $\zeta = 20$ *in Eqs.* (1.52) *and* (1.53).

For nonrelativistic scattering, we know from Eq. (1.53) of the text that

$$\cos\theta_{\rm Lab} = \frac{\cos\theta_{\rm CM}+\zeta}{(1+2\zeta\cos\theta_{\rm CM}+\zeta^2)^{1/2}}, \quad \zeta = \frac{m_1}{m_2}, \quad (1.14)$$

where m_1, m_2 represent respectively the masses of the projectile and the target in the laboratory frame. The first case that we want to consider, namely,

$$\zeta = \frac{m_1}{m_2} = 0.05 \quad \text{or} \quad m_2 = 20m_1, \quad (1.15)$$

corresponds to the scattering of a light projectile from a heavy target, while the second case

$$\zeta = \frac{m_1}{m_2} = 20 \quad \text{or} \quad m_1 = 20m_2, \quad (1.16)$$

describes the opposite, where a heavy projectile is scattered from a light target. The two cases, therefore, represent inverse scenarios. We will treat them separately. Furthermore, the solutions for the two cases can be worked out in two simple but equivalent ways as follows.

$\underline{\zeta = 0.05}$

Method I. Here we simply evaluate the angles numerically from the formula

$$\cos\theta_{\rm Lab} = \frac{\cos\theta_{\rm CM} + 0.05}{(1 + 0.1\cos\theta_{\rm CM} + 0.0025)^{1/2}}. \tag{1.17}$$

Keeping terms up to three-digit accuracy, we have

$\cos\theta_{\rm CM}$	$\cos\theta_{\rm Lab}$
−1.00	−1.000
−0.70	−0.673
−0.50	−0.461
−0.05	0.000
0.00	0.050
0.50	0.536
0.70	0.724
1.00	1.000

This is plotted in Fig. 1.1, and it is clear that for this case of a light projectile scattering from a heavy target, the scattering angles in the laboratory and in the center-of-mass frames are approximately the same.

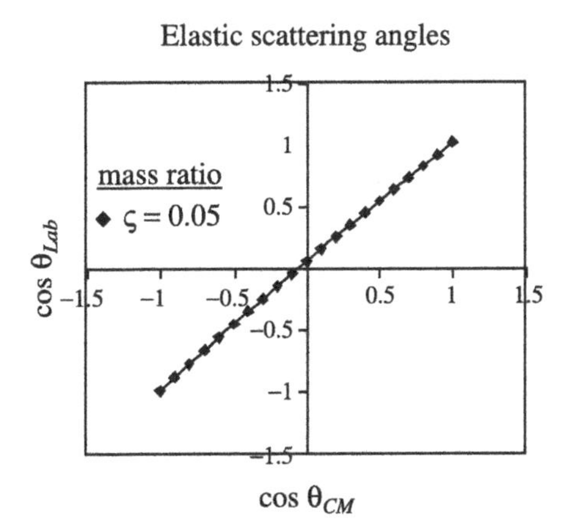

Fig. 1.1. Center-of-mass vs. laboratory scattering angles for a mass ratio $\zeta = 0.05$.

Method II. When $\zeta \ll 1$, as is true in the present case, we can write

$$\begin{aligned}\cos\theta_{\text{Lab}} &= \frac{\cos\theta_{\text{CM}} + \zeta}{(1 + 2\zeta\cos\theta_{\text{CM}} + \zeta^2)^{1/2}} \\ &\approx \frac{\cos\theta_{\text{CM}} + \zeta}{(1 + 2\zeta\cos\theta_{\text{CM}})^{1/2}} = (\cos\theta_{\text{CM}} + \zeta)(1 + 2\zeta\cos\theta_{\text{CM}})^{-1/2} \\ &\approx (\cos\theta_{\text{CM}} + \zeta)(1 - \zeta\cos\theta_{\text{CM}}) \\ &\approx \cos\theta_{\text{CM}} + \zeta(1 - \cos^2\theta_{\text{CM}}) \\ &= \cos\theta_{\text{CM}} + \zeta\sin^2\theta_{\text{CM}} = \cos\theta_{\text{CM}} + 0.05\sin^2\theta_{\text{CM}}. \end{aligned} \tag{1.18}$$

Here we have neglected terms of order ζ^2 and higher, which would lead to small corrections in the result. We note that since $0 \leq \sin^2\theta_{\text{CM}} \leq 1$, it follows that

$$\cos\theta_{\text{Lab}} \approx \cos\theta_{\text{CM}}, \tag{1.19}$$

as we saw from the explicit calculation.

<u>$\zeta = 20$</u>

Method I. In this case, a heavy projectile scatters from a light target and we have

$$\cos\theta_{\text{Lab}} = \frac{\cos\theta_{\text{CM}} + 20}{(1 + 40\cos\theta_{\text{CM}} + 400)^{1/2}}. \tag{1.20}$$

Explicit numerical evaluation, keeping terms up to four digit accuracy, leads to:

$\cos\theta_{\text{CM}}$	$\cos\theta_{\text{Lab}}$
−1.0	1.0000
−0.7	0.9993
−0.5	0.9990
0.0	0.9987
0.5	0.9991
0.7	0.9994
1.0	1.0000

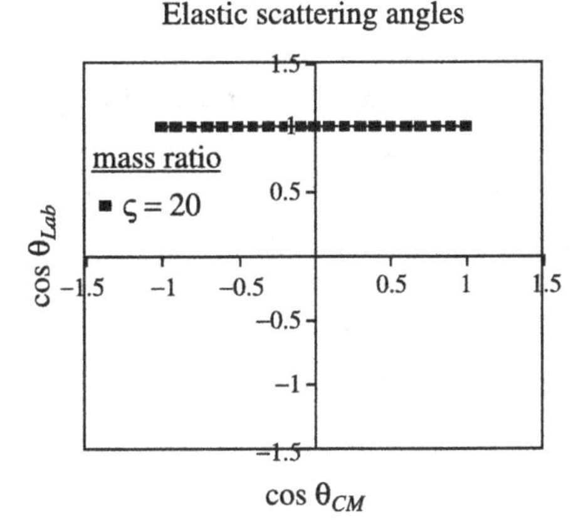

Fig. 1.2. Center-of-mass vs. laboratory scattering angles for a mass ratio $\zeta = 20$.

This is plotted in Fig. 1.2, and shows that the scattering in the laboratory frame is almost entirely in the forward direction — like a truck hitting a ping-pong ball.

Method II. When $\zeta \gg 1$, as is true in the present case, we can write

$$\begin{aligned}
\cos\theta_{\text{Lab}} &= \frac{\cos\theta_{\text{CM}} + \zeta}{(1 + 2\zeta\cos\theta_{\text{CM}} + \zeta^2)^{1/2}} \\
&= \frac{1}{\zeta}(\cos\theta_{\text{CM}} + \zeta)\left(1 + \frac{2}{\zeta}\cos\theta_{\text{CM}} + \frac{1}{\zeta^2}\right)^{-1/2} \\
&\approx \frac{1}{\zeta}(\cos\theta_{\text{CM}} + \zeta)\left(1 - \frac{1}{\zeta}\cos\theta_{\text{CM}} - \frac{1}{2\zeta^2} + \frac{3}{2\zeta^2}\cos^2\theta_{\text{CM}}\right) \\
&\approx \frac{1}{\zeta}\left(\zeta - \cos\theta_{\text{CM}} + \cos\theta_{\text{CM}} - \frac{1}{\zeta}\cos^2\theta_{\text{CM}} - \frac{1}{2\zeta} + \frac{3}{2\zeta}\cos^2\theta_{\text{CM}}\right) \\
&= 1 - \frac{1}{2\zeta^2}\sin^2\theta_{\text{CM}} = 1 - \frac{1}{800}\sin^2\theta_{\text{CM}}.
\end{aligned} \tag{1.21}$$

Here we have neglected higher-order terms in $\frac{1}{\zeta}$, which are negligible. From the fact that $0 \leq \sin^2\theta_{\text{CM}} \leq 1$, we conclude that in this case

$$\cos\theta_{\text{Lab}} \approx 1, \tag{1.22}$$

which is consistent with the numerical calculation. In either case (whether $\zeta = 0.05$ or $\zeta = 20$), we see directly from the tables that $\theta_{\text{Lab}} \leq \theta_{\text{CM}}$.

Problem 1.4 *What would be the approximate counting rate observed in the Rutherford scattering of* 10 MeV *α-particles off lead foil at an angle of* $\theta = \frac{\pi}{2}$ *in the laboratory? Assume an incident flux of* 10^6 *α-particles per second on the foil, a foil* 0.1 cm *thick, and a detector of transverse area* 1 cm × 1 cm *placed* 100 cm *from the interaction point, and density of lead of* 11.3 g/cm³. *What would be the counting rate at* $\theta = 5°$? *By about how much would your answers change if the above angles were specified for the center-of-mass — be quantitative, but use approximations where necessary. (Why don't you have to know the area of the foil?)*

From Eq. (1.40) of the text, the counting rate is given by

$$\mathrm{d}n(\theta) = N_0 \frac{\rho t A_0}{A} \frac{\mathrm{d}\sigma}{\mathrm{d}\Omega}(\theta)\mathrm{d}\Omega. \tag{1.23}$$

In the present problem of the scattering of α particles from a foil of lead ($^{208}\mathrm{Pb}^{82}$), we are given

$$\begin{aligned}
N_0 &= \text{incident flux/foil area} = 10^6\ \text{sec}^{-1}/\text{foil area}, \\
A_0 &= \text{Avogadro's number} = 6 \times 10^{23}/\text{mole}, \\
\rho &= \text{density of the foil} = 11.3\,\text{g/cm}^3, \\
t &= \text{thickness of the foil} = 0.1\,\text{cm}, \\
A &= \text{Atomic weight of lead} = 208, \\
E &= \text{energy of the incident } \alpha \text{ particle} = 10\,\text{MeV}.
\end{aligned} \tag{1.24}$$

Furthermore, we are also given that the detector has an area

$$\mathrm{d}s = 1\,\text{cm} \times 1\,\text{cm} = 1\,\text{cm}^2 \tag{1.25}$$

and is located at a distance

$$R = 100\,\text{cm}, \tag{1.26}$$

from the target (foil). Therefore, the solid angle subtended by the detector at the scattering center is given by

$$\mathrm{d}\Omega = \frac{\mathrm{d}s}{R^2} = 10^{-4}\ \text{sr}. \tag{1.27}$$

Finally, we note that for the scattering of α particles from lead ($^{208}\mathrm{Pb}^{82}$), we have

$$Z = 2, \quad Z' = 82, \tag{1.28}$$

so that the Rutherford scattering cross section takes the form

$$\begin{aligned}\frac{\mathrm{d}\sigma}{\mathrm{d}\Omega}(\theta) &= \left(\frac{ZZ'e^2}{4E}\right)^2 \left(\frac{1}{\sin\frac{\theta}{2}}\right)^4 (\mathrm{sr})^{-1}\\ &= \left(2\times 82\times \frac{\hbar c}{4\times 10\,\mathrm{MeV}}\times\frac{e^2}{\hbar c}\right)^2 \operatorname{cosec}^4\frac{\theta}{2}(\mathrm{sr})^{-1}, \qquad (1.29)\end{aligned}$$

where we note that

$$\hbar c \approx 197\ \mathrm{MeV}-\mathrm{F},\quad \frac{e^2}{\hbar c} = \text{fine structure constant} = \frac{1}{137}, \qquad (1.30)$$

and 1 F = 1 fm or 1 Fermi = 10^{-13} cm. Using these values, the differential cross section at any angle takes the form

$$\begin{aligned}\frac{\mathrm{d}\sigma}{\mathrm{d}\Omega}(\theta) &= \left(164\times\frac{197\ \mathrm{MeV}-\mathrm{F}}{40\ \mathrm{MeV}}\times\frac{1}{137}\right)^2 \operatorname{cosec}^4\frac{\theta}{2}(\mathrm{sr})^{-1}\\ &\approx 0.4\times 10^{-24}\operatorname{cosec}^4\frac{\theta}{2}\,\mathrm{cm}^2/\mathrm{sr}. \qquad (1.31)\end{aligned}$$

Using all of these results, we can calculate the counting rate at any angle from (1.23) as:

$$\begin{aligned}\mathrm{d}n(\theta) &= 10^6/\mathrm{sec}\times\frac{6\times 10^{23}\times 11.3\times 0.1/\mathrm{cm}^2}{208}\\ &\quad\times 0.4\times 10^{-24}\operatorname{cosec}^4\frac{\theta}{2}\,\mathrm{cm}^2/\mathrm{sr}\times 10^{-4}\,\mathrm{sr}\\ &\approx 0.13\operatorname{cosec}^4\frac{\theta}{2}\ \mathrm{counts/sec}. \qquad (1.32)\end{aligned}$$

It follows now that

$$\boxed{\mathrm{d}n(\theta=\tfrac{\pi}{2}) \approx 0.13\times 4\,\mathrm{counts/sec} \approx 0.5\,\mathrm{counts/sec},}$$

$$\boxed{\mathrm{d}n(\theta=5^\circ=\tfrac{\pi}{36}) \approx 0.13\times 28\times 10^4\ \mathrm{counts/sec} \approx 3.6\times 10^4\ \mathrm{counts/sec}.}$$

For α particles scattering from lead ($^{208}\mathrm{Pb}^{82}$), we have $\zeta = \frac{m_1}{m_2} = \frac{4}{208} \approx 0.02$. As a result, using (1.18) we obtain

$$\cos\theta_L = \cos\theta_{\mathrm{CM}} + \zeta\sin^2\theta_{\mathrm{CM}} = \cos\theta_{\mathrm{CM}} + 0.02\sin^2\theta_{\mathrm{CM}}, \tag{1.33}$$

so that for $\theta_{\mathrm{CM}} = \frac{\pi}{2}$, we obtain

$$\cos\theta_{\mathrm{Lab}} = 0.02$$
$$\text{or}\quad \theta_{\mathrm{Lab}} \approx \frac{\pi}{2} - 0.02,$$

which leads to

$$\Delta\theta|_{\theta_{\mathrm{CM}}=\frac{\pi}{2}} = \theta_{\mathrm{CM}} - \theta_{\mathrm{Lab}} \approx 0.02. \tag{1.34}$$

On the other hand, for $\theta_{\mathrm{CM}} = 5° = \frac{\pi}{36}$, we have

$$\begin{aligned}\cos\theta_{\mathrm{Lab}} &= \cos\theta_{\mathrm{CM}} + 0.02\sin^2\theta_{\mathrm{CM}} \approx 1 - \frac{1}{2}\left(\frac{\pi}{36}\right)^2 + 0.02\left(\frac{\pi}{36}\right)^2\\ &= 1 - \frac{1}{2}\times 0.96\left(\frac{\pi}{36}\right)^2 \approx 1 - \frac{1}{2}\left(0.98\times\frac{\pi}{36}\right)^2\\ \text{or}\quad \theta_{\mathrm{Lab}} &\approx 0.98\times\frac{\pi}{36},\end{aligned}$$

which leads to

$$\Delta\theta|_{\theta_{\mathrm{CM}}=5°=\frac{\pi}{36}} = \theta_{\mathrm{CM}} - \theta_{\mathrm{Lab}} \approx 0.02\times\frac{\pi}{36} \approx 0.002. \tag{1.35}$$

From Eq. (1.32), the relative change in counting rate arising from this difference in angle is given approximately by:

$$\frac{|\Delta(\mathrm{d}n)|}{\mathrm{d}n}(\theta) = \frac{4\frac{\Delta\theta}{2}\cos\frac{\theta}{2}}{(\sin\frac{\theta}{2})^5}\bigg/\frac{1}{(\sin\frac{\theta}{2})^4} = 2\Delta\theta\cot\frac{\theta}{2}. \tag{1.36}$$

This leads to

$$\boxed{\frac{|\Delta(\mathrm{d}n)|}{\mathrm{d}n}(\theta = \tfrac{\pi}{2}) \approx 2\times 0.02\times 1 = 0.04 = 4\%,}$$

$$\boxed{\frac{|\Delta(\mathrm{d}n)|}{\mathrm{d}n}(\theta = \tfrac{\pi}{36}) \approx 2\times 0.002\times 23 \approx 0.092 = 9.2\%.}$$

Problem 1.5 *Sketch the cross section in the laboratory frame as a function of* $\cos\theta_{\text{Lab}}$ *for the elastic scattering of equal-mass particles when* $\frac{d\sigma}{d\Omega_{\text{CM}}}$ *is isotropic and equal to* 100 mb/sr. *What would be your result for* $\zeta = 0.05$ *in Eq.* (1.52)? (*You may use approximations where necessary.*)

We know from Eqs. (1.53) and (1.55) of the text that

$$\cos\theta_{\text{Lab}} = \frac{\cos\theta_{\text{CM}} + \zeta}{(1 + 2\zeta\cos\theta_{\text{CM}} + \zeta^2)^{1/2}},$$
$$\frac{d\sigma}{d\Omega_{\text{Lab}}}(\theta_{\text{Lab}}) = \frac{d\sigma}{d\Omega_{\text{CM}}}\,\frac{(1 + 2\zeta\cos\theta_{\text{CM}} + \zeta^2)^{3/2}}{|1 + \zeta\cos\theta_{\text{CM}}|}, \tag{1.37}$$

where $\zeta = \frac{m_1}{m_2}$.

When $m_1 = m_2$, namely, the particle masses are equal, we have $\zeta = 1$, and we obtain from the first relationship

$$\cos\theta_{\text{Lab}} = \frac{1 + \cos\theta_{\text{CM}}}{\sqrt{2}(1 + \cos\theta_{\text{CM}})^{1/2}} = \left(\frac{1 + \cos\theta_{\text{CM}}}{2}\right)^{1/2},$$

which leads to

$$\begin{aligned}\frac{(1 + 2\zeta\cos\theta_{\text{CM}} + \zeta^2)^{3/2}}{|1 + \zeta\cos\theta_{\text{CM}}|} &= \frac{2^{3/2}(1 + \cos\theta_{\text{CM}})^{3/2}}{(1 + \cos\theta_{\text{CM}})} \\ &= 2^{3/2}\,(1 + \cos\theta_{\text{CM}})^{1/2} \\ &= 4\left(\frac{1 + \cos\theta_{\text{CM}}}{2}\right)^{1/2} \\ &= 4\cos\theta_{\text{Lab}}.\end{aligned} \tag{1.38}$$

Thus, for $\zeta = 1$ (equal masses), we can write

$$\frac{d\sigma}{d\Omega_{\text{Lab}}}(\theta_{\text{Lab}}) = 4\frac{d\sigma}{d\Omega_{\text{CM}}}(\theta_{\text{CM}})\cos\theta_{\text{Lab}}. \tag{1.39}$$

For an isotropic cross section in the center-of-mass frame of value

$$\frac{d\sigma}{d\Omega_{\text{CM}}}(\theta_{\text{CM}}) = 100\,\text{mb/sr}, \tag{1.40}$$

we obtain

$$\frac{d\sigma}{d\Omega_{\text{Lab}}}(\theta_{\text{Lab}}) = 400\cos\theta_{\text{Lab}}\,\text{mb/sr}, \tag{1.41}$$

which leads to a straight line with zero intercept for the laboratory cross section plotted against $\cos\theta_{\text{Lab}}$ (see Fig. 1.3).

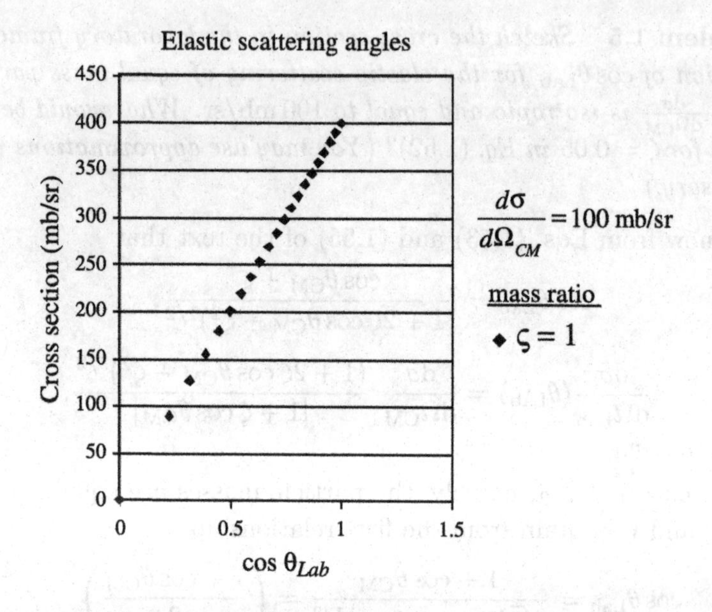

Fig. 1.3. Differential cross section vs. laboratory angle for mass ratio $\zeta = 1.0$.

For the case of a light projectile with $\zeta = 0.05$, we can use our earlier result (1.18) to write

$$\cos\theta_{\rm Lab} \approx \cos\theta_{\rm CM} + 0.05\sin^2\theta_{\rm CM}, \tag{1.42}$$

which can be inverted to give

$$\cos\theta_{\rm CM} \approx \cos\theta_{\rm Lab} - 0.05. \tag{1.43}$$

Using this, we obtain

$$\begin{aligned}\frac{(1+2\zeta\cos\theta_{\rm CM}+\zeta^2)^{3/2}}{|1+\zeta\cos\theta_{\rm CM}|} &\approx \frac{(1+0.1\cos\theta_{\rm CM})^{3/2}}{(1+0.05\cos\theta_{\rm CM})}\\ &\approx (1+0.15\cos\theta_{\rm CM})(1-0.05\cos\theta_{\rm CM})\\ &\approx 1+0.1\cos\theta_{\rm CM}\\ &\approx 1+0.1(\cos\theta_{\rm Lab}-0.05)\\ &\approx 1+0.1\cos\theta_{\rm Lab}.\end{aligned} \tag{1.44}$$

Therefore, for an isotropic cross section in the center-of-mass given by (1.40), the cross section in the laboratory takes the form

$$\frac{\mathrm{d}\sigma}{\mathrm{d}\Omega_{\mathrm{Lab}}}(\theta_{\mathrm{Lab}}) \approx \frac{\mathrm{d}\sigma}{\mathrm{d}\Omega_{\mathrm{CM}}}(\theta)(1 + 0.1\cos\theta_{\mathrm{Lab}}) = 100(1 + 0.1\cos\theta_{\mathrm{Lab}})$$
$$= (100 + 10\cos\theta_{\mathrm{Lab}})\,\mathrm{mb/sr}. \tag{1.45}$$

In this case, the laboratory cross section is again linear with $\cos\theta_{\mathrm{Lab}}$, but has a much smaller slope and a finite intercept, as shown in Fig. 1.4.

Note that $\int_{\mathrm{CM}} d\sigma = \int_{\mathrm{Lab}} d\sigma$!

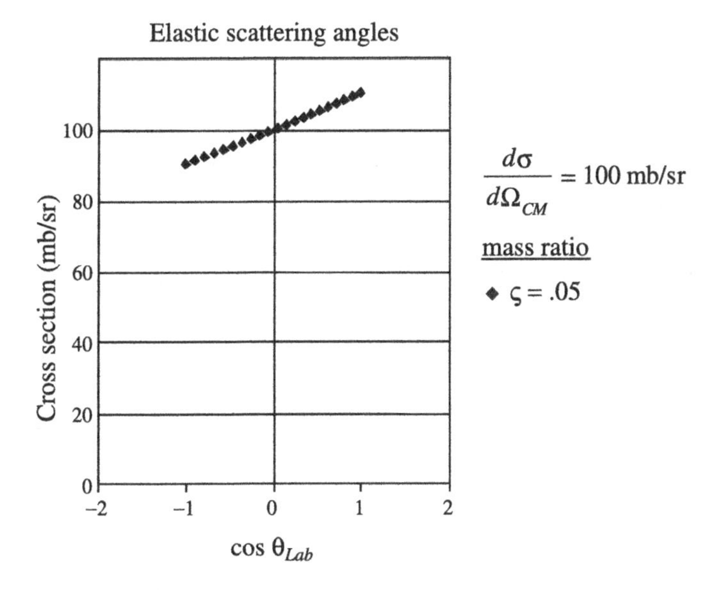

Fig. 1.4. Differential cross section vs. laboratory angle for mass ratio $\zeta = 0.05$.

Problem 1.6 *Certain radioactive nuclei emit α particles. If the kinetic energy of these α particles is 4 MeV, what is their velocity if you assume them to be nonrelativistic? How large an error do you make in neglecting special relativity in the calculation of v? What is the closest that such an α particle can get to the center of a Au nucleus?*

The α particle has a rest mass given approximately by

$$M = 4 \times 10^3 \,\text{MeV}/c^2. \tag{1.46}$$

If the α particle has a kinetic energy

$$T = 4\,\text{MeV}, \tag{1.47}$$

and if we treat it nonrelativistically, then we have

$$\begin{aligned} T = \frac{1}{2}Mv_{\text{NR}}^2 = \frac{1}{2}Mc^2 \times \left(\frac{v_{\text{NR}}}{c}\right)^2 &= 4\text{ MeV} \\ \text{or}\quad \frac{1}{2} \times 4 \times 10^3 \text{ MeV} \times \left(\frac{v_{\text{NR}}}{c}\right)^2 &= 4\text{ MeV} \\ \text{or}\quad \left(\frac{v_{\text{NR}}}{c}\right)^2 &= 2 \times 10^{-3} \\ \text{or}\quad \frac{v_{\text{NR}}}{c} &= \sqrt{20} \times 10^{-2} \approx 0.045. \end{aligned} \tag{1.48}$$

Here v_{NR} represents the magnitude of the velocity of the nonrelativistic particle.

On the other hand, if we treat the α particle as relativistic, we can then use the relativistic relationships from Eqs. (A.7) and (A.10) of Appendix A of the text to write

$$\begin{aligned} E &= \gamma Mc^2, \\ T &= E - Mc^2 = (\gamma - 1)Mc^2, \\ cP &= \sqrt{T^2 + 2Mc^2T} = \sqrt{\gamma^2 - 1}Mc^2, \end{aligned} \tag{1.49}$$

where P is the magnitude of the momentum. The relativistic velocity now follows using Eq. (A.8) of the text

$$\frac{v_{\text{R}}}{c} = \frac{cP}{E} = \frac{\sqrt{(\gamma + 1)(\gamma - 1)}}{\gamma}. \tag{1.50}$$

From the fact that the α particle has kinetic energy

$$T = 4\,\text{MeV}, \tag{1.51}$$

we can determine the Lorentz factor using (1.49)

$$\begin{aligned} T = (\gamma - 1)Mc^2 &= 4\,\text{MeV} \\ \text{or}\quad (\gamma - 1) \times 4 \times 10^3\,\text{MeV} &= 4\,\text{MeV} \\ \text{or}\quad \gamma &= 1 + 10^{-3}. \end{aligned} \tag{1.52}$$

Using this in (1.50), we can determine the relativistic velocity

$$\begin{aligned}\frac{v_{\mathrm{R}}}{c} &= \frac{\sqrt{(2+10^{-3})10^{-3}}}{1+10^{-3}} \\ &\approx \sqrt{20}\times 10^{-2}(1+0.5\times 10^{-3})^{1/2}(1-10^{-3}) \\ &\approx \sqrt{20}\times 10^{-2}(1+0.025\times 10^{-3})(1-10^{-3}) \\ &\approx \sqrt{20}\times 10^{-2}(1-0.00075).\end{aligned} \tag{1.53}$$

Using (1.48) we see that we can write the relativistic velocity in Eq. (1.53) as

$$v_{\mathrm{R}} = v_{\mathrm{NR}}(1-0.00075). \tag{1.54}$$

Consequently, we can define the relative error in neglecting relativity as

$$\frac{|\Delta v|}{v_{\mathrm{NR}}} = \frac{|v_{\mathrm{R}} - v_{\mathrm{NR}}|}{v_{\mathrm{NR}}} \approx 0.00075 = 0.07\%. \tag{1.55}$$

For the scattering of such an α particle from gold (Au), the distance of closest approach can be dertermined as follows. First we note that the distance of closest approach is attained when the impact parameter vanishes (for head on collisions). From Eq. (1.25) of the text, we see that the distance of closest in this case ($b = 0$) is given by

$$r_0 = \frac{ZZ'e^2}{E}, \tag{1.56}$$

where, for scattering of α particles from gold (Au), we have

$$Z = 2, \quad Z' = 79. \tag{1.57}$$

If we treat the α particle nonrelativistically, we have

$$E = T = 4\,\mathrm{MeV}. \tag{1.58}$$

Using all of these, we determine

$$\begin{aligned}r_0 &= ZZ' \times \frac{\hbar c}{E} \times \frac{e^2}{\hbar c} \\ &= 2\times 79 \times \frac{197\ \mathrm{MeV} - \mathrm{F}}{4\ \mathrm{MeV}} \times \frac{1}{137} \\ &\approx 56\ \mathrm{F} = 5.6\times 10^{-12}\,\mathrm{cm}.\end{aligned} \tag{1.59}$$

Problem 1.7 *An electron of momentum* 0.511 MeV/c *is observed in the laboratory. What are its* $\beta = \frac{v}{c}$, $\gamma = (1-\beta^2)^{-1/2}$, *kinetic energy, and total energy?*

The rest mass of an electron is

$$m = 0.511\,\text{MeV}/c^2. \tag{1.60}$$

If the electron has a momentum

$$p = 0.511\,\text{MeV}/c, \tag{1.61}$$

then it is fairly relativistic. Using Einstein's relationship we have the total energy of the electron

$$E = \sqrt{p^2c^2 + m^2c^4} = \sqrt{2}\,mc^2 = \gamma mc^2, \tag{1.62}$$

where we have used (1.49). This determines the Lorentz factor

$$\gamma = \sqrt{2}. \tag{1.63}$$

From the definition of the Lorentz factor

$$\gamma = \frac{1}{(1-\beta^2)^{1/2}}, \quad \beta = \frac{v}{c}, \tag{1.64}$$

we obtain

$$\beta = \left(1 - \frac{1}{\gamma^2}\right)^{1/2} = \left(1 - \frac{1}{2}\right)^{1/2} = \frac{1}{\sqrt{2}}. \tag{1.65}$$

The value of the total energy is obtained from (1.62)

$$E = \sqrt{2}mc^2 = \sqrt{2} \times 0.511\text{ MeV} \approx 0.722\,\text{MeV}. \tag{1.66}$$

The kinetic energy follows from (1.49)

$$T = (\gamma - 1)mc^2 = (\sqrt{2} - 1) \times 0.511\text{ MeV} \approx 0.211\,\text{MeV}. \tag{1.67}$$

Problem 1.8 *What are the approximate values of the kinetic energy for the recoiling lead nucleus and the momentum transfers (in* eV *units) at the cutoffs specified in Problem* 1.1*?*

We note from Eq. (1.1) of the text that in a scattering involving α particles, conservation of momentum determines the recoil velocity of the target

$$\vec{v}_t = \frac{m_\alpha}{m_t}(\vec{v}_0 - \vec{v}_\alpha), \tag{1.68}$$

where m_α, m_t represent the mass of the α particle and target, respectively, and $\vec{v}_0$, $\vec{v}_\alpha$ denote respectively the initial and final velocities of the α particle in the laboratory. It follows that the recoil kinetic energy of the target in the laboratory is given by

$$\begin{aligned}
\text{Recoil energy} &= \frac{1}{2}m_t v_t^2 = \frac{m_t}{2} \times \frac{m_\alpha^2}{m_t^2}(\vec{v}_0 - \vec{v}_\alpha)^2 \\
&= \frac{1}{2m_t}(\vec{p}_0 - \vec{p}_\alpha)^2 \\
&= \frac{1}{2m_t}\left(p_0^2 + p_\alpha^2 - 2\vec{p}_0 \cdot \vec{p}_\alpha\right). \qquad (1.69)
\end{aligned}$$

For elastic scattering, because m_α is far smaller than m_t, the magnitude of the initial and final momentum of the α particle is essentially the same. (The nucleus absorbs essentially no energy, but just momentum.) We can therefore write

$$\begin{aligned}
\text{Recoil energy} &\approx \frac{1}{m_t}p_0^2(1 - \cos\theta) \\
&= \frac{2m_\alpha}{m_t}E(1 - \cos\theta) = \frac{4m_\alpha}{m_t}E\sin^2\frac{\theta}{2} \\
&= \frac{4m_\alpha}{m_t}\frac{E}{1 + b^2\left(\frac{2E}{ZZ'e^2}\right)^2}, \qquad (1.70)
\end{aligned}$$

where E represents the energy of the incident α particle, and we used the relationship between the scattering angle and the impact parameter given in Eq. (1.32) of the text. For the scattering of the α particle from lead ($^{208}\text{Pb}^{82}$), as in Eq. (1.31), we can calculate

$$\begin{aligned}
\left(\frac{ZZ'e^2}{2E}\right)^2 &= \left(ZZ' \times \frac{\hbar c}{2E} \times \frac{e^2}{\hbar c}\right)^2 \\
&= \left(2 \times 82 \times \frac{197\ \text{MeV} - \text{F}}{2 \times 10\ \text{MeV}} \times \frac{1}{137}\right)^2 \\
&\approx 1.4 \times 10^{-24}\ \text{cm}^2. \qquad (1.71)
\end{aligned}$$

Using this, we can write the recoil energy as a function of impact parameter as

$$\begin{aligned}\text{Recoil energy} &= \frac{4m_\alpha}{m_t}\,\frac{E}{1+b^2\left(\frac{2E}{ZZ'e^2}\right)^2}\\ &\approx \frac{4\times 4}{208}\,\frac{10\text{ MeV}}{1+\frac{b^2}{1.4\times 10^{-24}\text{ cm}^2}}\\ &= \frac{0.8\,\text{MeV}}{1+0.7\times 10^{24}\,b^2/\text{cm}^2}. \end{aligned} \tag{1.72}$$

Using Eq. (1.72), we tabulate below the values of recoil energy for different cutoffs on impact parameter:

b (cm)	Recoil energy (MeV)
10^{-12}	0.5
10^{-10}	1.1×10^{-4}
10^{-8}	1.1×10^{-8}

Since $T_{\text{Pb}} \ll T_\alpha$, our assumption that the initial and final energies of the α particle are the same holds well. Note that, for impact parameters of 10^{-8} cm, energy transfers to the nucleus are vanishingly small 10^{-2} eV. At these "enormous" distances, there must be some shielding of nuclear charge by the external electrons, and so the calculation cannot be valid. Also, for transferring energy to bound electrons, rather than to nuclei, the electrons cannot absorb arbitrary amounts since they are located in quantized orbits.

Problem 1.9 *Taking the ultrarelativistic limit of Eq. (1.71), find an approximate expression for θ_{Lab} at $\theta_{\text{CM}} = \frac{\pi}{2}$, and evaluate θ_{Lab} for $\gamma_{\text{CM}} = 10$ and $\gamma_{\text{CM}} = 100$. Does the approximation hold best for particles with small or large mass values?*

Equation (1.71) in the text relates the laboratory and CM angles as follows:

$$\tan\theta_{\text{Lab}} = \frac{1}{\gamma_{\text{CM}}}\,\frac{\tilde{\beta}\sin\theta_{\text{CM}}}{\tilde{\beta}\cos\theta_{\text{CM}}+\beta_{\text{CM}}} \approx \frac{1}{\gamma_{\text{CM}}}\,\frac{\sin\theta_{\text{CM}}}{\cos\theta_{\text{CM}}+1} \approx \frac{1}{\gamma_{\text{CM}}} \tag{1.73}$$

where the next-to-last term is for highly relativistic scattering, and the final term is for $\theta_{\text{CM}} = \frac{\pi}{2}$.

Thus $\theta_{\text{Lab}} = \arctan\left(\frac{1}{\gamma_{\text{CM}}}\right)$. For $\gamma_{\text{CM}} = 10$ and 100, this corresponds to $\arctan(0.1) = 5.7^\circ$ and $\arctan(0.01) = 0.57^\circ$, respectively.

Clearly, such approximations hold best at high energies when both β_{CM} and the $\tilde{\beta}$ of the produced particle are large. That is, when the particle mass in $\tilde{\beta}$ can be ignored. It should be recognized that high-energy scattering does not necessarily imply that the mass of any produced particle can be assumed to be negligible.

Problem 1.10 *What is the minimum impact parameter needed to deflect* 7.7 MeV *α-particles from gold nuclei by at least* 1°? *What about by at least* 30°? *What is the ratio of probabilities for deflections of* $\theta > 1^\circ$ *relative to* $\theta > 30^\circ$? *(See the CRC Handbook for the density of gold.)*

For the scattering of a 7.7 MeV α-particle from gold, we have

$$Z = 2, \quad Z' = 79, \quad E = 7.7\,\text{MeV}, \tag{1.74}$$

so that we obtain

$$\begin{aligned} \frac{ZZ'e^2}{2E} &= ZZ' \times \frac{\hbar c}{2E} \times \frac{e^2}{\hbar c} \\ &= 2 \times 79 \times \frac{197\,\text{MeV} - \text{F}}{2 \times 7.7\,\text{MeV}} \times \frac{1}{137} \\ &\approx 14.5 \times 10^{-13}\,\text{cm} \approx 1.4 \times 10^{-12}\,\text{cm}. \end{aligned} \tag{1.75}$$

We know from Eq. (1.32) of the text that

$$b = \frac{ZZ'e^2}{2E} \cot\frac{\theta}{2}, \tag{1.76}$$

which leads to

$$b(\theta) \approx 1.4 \times 10^{-12} \cot\frac{\theta}{2}\,\text{cm}. \tag{1.77}$$

We note that for

$$\theta = 1^\circ = \frac{\pi}{180} \approx \frac{1}{60} \ll 1, \tag{1.78}$$

we have

$$\tan\frac{\theta}{2} \approx \frac{\theta}{2} \approx \frac{1}{120}, \quad \cot\frac{\theta}{2} = \frac{1}{\tan\frac{\theta}{2}} \approx 120. \tag{1.79}$$

Similarly, for

$$\theta = 30^\circ = \frac{\pi}{6} \approx \frac{1}{2} \tag{1.80}$$

we have

$$\tan\frac{\theta}{2} \approx \frac{\theta}{2} \approx \frac{1}{4}, \quad \cot\frac{\theta}{2} = \frac{1}{\tan\frac{\theta}{2}} \approx 4. \tag{1.81}$$

Using these values in (1.77), we obtain

$$\begin{aligned} b(\theta = 1^\circ) &\approx 1.4 \times 10^{-12} \times 120\,\text{cm} = 1.7 \times 10^{-10}\,\text{cm}, \\ b(\theta = 30^\circ) &\approx 1.4 \times 10^{-12} \times 4\,\text{cm} = 5.6 \times 10^{-12}\,\text{cm}. \end{aligned} \tag{1.82}$$

As we have already seen in Problem 1.1, the probability of scattering for angles greater than θ_b goes as the area πb^2. Therefore, using (1.82) we have

$$\begin{aligned} \frac{\sigma(\theta > 1^\circ)}{\sigma(\theta > 30^\circ)} &= \frac{b^2(\theta = 1^\circ)}{b^2(\theta = 30^\circ)} \\ &\approx \left(\frac{1.7 \times 10^{-10}}{5.6 \times 10^{-12}}\right)^2 \approx 900. \end{aligned} \tag{1.83}$$

In other words, there will be approximately 900 more particle collisions for $\theta > 1^\circ$ than for $\theta > 30^\circ$.

Problem 1.11 *Consider a collimated source of* 8 MeV *α-particles that provides* 10^4 α/sec *that impinge on a* 0.1 mm *gold foil. What counting rate would you expect in a detector that subtends an annular cone of* $\Delta\theta = 0.05$ *rad, at a scattering angle of* $\theta = 90^\circ$*? Compare this to the rate at* $\theta = 5^\circ$. *Is there a problem? Is it serious (see Problem 1.12). (Hint: You can use the small-angle approximation where appropriate, and find the density of gold in the CRC Handbook.)*

For scattering of 8 MeV α particles from gold, we have

$$\begin{aligned}
Z &= 2, \quad Z' = 79, \quad E = 8\,\text{MeV},\\
\rho &= \text{density of gold} \approx 19.3\,\text{g/cm}^3,\\
t &= \text{Thickness of gold foil} = 0.1\,\text{mm} = 10^{-2}\,\text{cm},\\
N_0 &= \text{Incident flux} = 10^4/\text{sec},\\
A &= \text{Atomic weight of gold} = 197,\\
A_0 &= \text{Avogadro's number} \approx 6\times 10^{23}/\text{mole},\\
\Delta\theta &= \text{Angle subtended by the detector} = 0.05\,\text{rad}.
\end{aligned} \tag{1.84}$$

We can therefore calculate

$$\begin{aligned}
\frac{\mathrm{d}\sigma}{\mathrm{d}\Omega}(\theta) &= \left(\frac{ZZ'e^2}{4E}\right)^2 \text{cosec}^4\frac{\theta}{2}(\text{sr})^{-1}\\
&= \left(ZZ'\times\frac{\hbar c}{4E}\times\frac{e^2}{\hbar c}\right)^2 \text{cosec}^4\frac{\theta}{2}(\text{sr})^{-1}\\
&= \left(2\times 79\times\frac{197\ \text{MeV}-\text{F}}{4\times 8\ \text{MeV}}\times\frac{1}{137}\right)^2 \text{cosec}^4\frac{\theta}{2}(\text{sr})^{-1}\\
&\approx 0.48\times 10^{-24}\ \text{cosec}^4\frac{\theta}{2}\ \text{cm}^2/\text{sr}.
\end{aligned} \tag{1.85}$$

Similarly, we have

$$\begin{aligned}
\frac{A_0\rho t}{A} &\approx \frac{6\times 10^{23}\times 19.3\times 10^{-2}/\text{cm}^2}{197}\\
&\approx 6\times 10^{20}/\text{cm}^2.
\end{aligned} \tag{1.86}$$

From Eq. (1.40) of the text we therefore obtain the counting rate

$$\begin{aligned}
\mathrm{d}n(\theta) &= N_0\frac{A_0\rho t}{A}\frac{\mathrm{d}\sigma}{\mathrm{d}\Omega}(\theta)\,\mathrm{d}\Omega\\
&\approx 10^4/\text{sec}\times 6\times 10^{20}/\text{cm}^2\times 0.48\times 10^{-24}\text{cosec}^4\frac{\theta}{2}\ \text{cm}^2/\text{sr}\times\mathrm{d}\Omega\\
&= 2.88\,\text{cosec}^4\frac{\theta}{2}\,\mathrm{d}\Omega(\text{sec}-\text{sr})^{-1}.
\end{aligned} \tag{1.87}$$

For scattering with azimuthal symmetry, we can write

$$\mathrm{d}\Omega = 2\pi\sin\theta\mathrm{d}\theta, \tag{1.88}$$

and if we identify $\mathrm{d}\theta \approx \Delta\theta = 0.05\,\text{rad}$, we get

$$\mathrm{d}\Omega \approx 2\pi\sin\theta\times 0.05\,\text{sr} \approx 0.3\sin\theta\,\text{sr}. \tag{1.89}$$

Putting this back into (1.87), we obtain

$$\begin{aligned} dn(\theta) &\approx 2.88 \text{ cosec}^4 \frac{\theta}{2} (\text{sec} - \text{sr})^{-1} \times 0.3 \sin\theta \text{ sr} \\ &\approx 0.86 \sin\theta \text{ cosec}^4 \frac{\theta}{2} (\text{sec})^{-1} . \end{aligned} \tag{1.90}$$

It follows that

$$\mathrm{d}n\left(\theta = \frac{\pi}{2}\right) \approx 0.86 \times 1 \times (\sqrt{2})^4 \approx 3.4/\text{sec}. \tag{1.91}$$

On the other hand, for $\theta = 5° = \frac{\pi}{36} \approx \frac{1}{12} \ll 1$, we have

$$\sin\theta \approx \theta \approx \frac{1}{12}, \quad \text{cosec}^4 \frac{\theta}{2} \approx \left(\frac{2}{\theta}\right)^4 \approx (24)^4, \tag{1.92}$$

and we obtain

$$\mathrm{d}n(\theta = 5°) \approx 0.86 \times \frac{1}{12} \times (24)^4/\text{sec} \approx 2.4 \times 10^4/\text{sec}. \tag{1.93}$$

This is, in fact, larger than the incident flux, and, if this were true, conservation of probability (particle number) would be violated, which is a serious problem! For one thing, we note that the approximation

$$\mathrm{d}\theta \approx \Delta\theta, \tag{1.94}$$

is meaningful only when

$$\frac{\Delta\theta}{\theta} \ll 1, \tag{1.95}$$

which is clearly violated when $\Delta\theta = 0.05$ rad and $\theta = 5° \approx \frac{1}{12} \approx 0.08$ rad. This is one of the sources of the difficulty. For other sources of this error, we turn to the solution of the next problem.

Problem 1.12 *Consider the expression Eq.* (1.41) *for Rutherford Scattering of α-particles from gold nuclei. Integrate this over all angles to obtain n. In principle, n cannot exceed N_0, the number of incident particles. Why? What cutoff value for θ would be required in the integral, that is, some $\theta = \theta_0 > 0$, to assure that n does not exceed N_0 in Problem 1.4? (Hint: After integrating, use the small-angle approximation to simplify the calculation.) Using the Heisenberg uncertainty principle $\Delta p_x \Delta x \approx \hbar$, where Δx is some* *transverse distance corresponding to a change in transverse momentum of* $\Delta p_x = p_{in}\theta_0 \approx \sqrt{2mE}\theta_0$*, calculate the distances* Δx *to which you have to restrict the description of the scattering. Are these distances sufficiently restrictive? Explain!*

Both Eqs. (1.40) and (1.41) of the text are equivalent and give the counting rate at a scattering angle θ. Let us look at Eq. (1.40) of the text

$$\frac{\mathrm{d}n}{\mathrm{d}\Omega}(\theta) = N_0 \frac{A_0 \rho t}{A} \frac{\mathrm{d}\sigma}{\mathrm{d}\Omega}(\theta). \tag{1.96}$$

Since N_0, A_0, ρ, t, A are constants independent of scattering angle, integrating the above relationship over all angles (above a certain cutoff value, corresponding to some cutoff in impact parameter, as was discussed in Problem 1.1), we obtain

$$\begin{aligned} n(b > b_0) &= N_0 \frac{A_0 \rho t}{A}\, \sigma(b > b_0) = N_0 \frac{A_0 \rho t}{A} \times \pi b_0^2 \\ \text{or} \quad \frac{n(b > b_0)}{N_0} &= \frac{A_0 \rho t}{A} \times \pi b_0^2. \end{aligned} \tag{1.97}$$

Clearly, the total number of particles scattered per second, $n(b > b_0)$, cannot exceed the total incident flux, namely, $n(b > b_0) \leq N_0$ for conservation of probability (particle number). This leads to the inequality

$$\begin{aligned} &\frac{A_0 \rho t}{A} \times \pi b_0^2 \leq 1 \\ &\text{or} \quad b_0 \leq \left(\frac{1}{\pi} \frac{A}{A_0 \rho t}\right)^{1/2}. \end{aligned} \tag{1.98}$$

For the case of the α-particle scattering from gold discussed in Problem 1.11, we can use (1.86) to obtain

$$b_0 \leq \left(\frac{1}{\pi} \frac{1}{6 \times 10^{20}}\right)^{1/2} \text{cm} \approx 2.2 \times 10^{-11}\,\text{cm}. \tag{1.99}$$

On the other hand, from Eq. (1.32) of the text

$$b = \frac{ZZ'e^2}{2E} \cot \frac{\theta_b}{2}, \tag{1.100}$$

we can calculate

$$\cot\frac{\theta_{b_0}}{2} = \frac{2E}{ZZ'e^2}\, b_0 = \frac{1}{ZZ'} \times \frac{2E}{\hbar c} \times \frac{\hbar c}{e^2}\, b_0$$

$$\leq \frac{1}{2\times 79} \times \frac{2\times 8\ \text{MeV}}{197\ \text{MeV} - \text{F}}$$

$$\times\, 137 \times 2.2 \times 10^{-11}\ \text{cm} \approx 16 \qquad (1.101)$$

$$\text{or} \quad \cot\frac{\theta_{b_0}}{2} \approx \frac{2}{\theta_{b_0}} \leq 16$$

$$\text{or} \quad \theta_{b_0} \geq \frac{2}{16} = \frac{1}{8} \approx \frac{\pi}{24} = 7.5^\circ.$$

For scattering angles below this bound, namely $\theta < 7.5^\circ$, conservation of probability will be violated, which is what we saw explicitly in the solution of Problem 1.11 for $\theta = 5^\circ$. Although derived from very physical considerations, this bound on the scattering angle seems artificial. After all, a particle can scatter at any angle, and should not be subject to any such bound. In fact, all of this is an artifact of our formalism, and can be seen as follows. First, we note that the bound on the impact parameter (1.99) is close to the size of the nucleus, and consequently imposing such a cutoff may affect the validity of the calculation. In fact, at these low energies the probability of scattering is very high, but in our derivation we assumed it to be low (no second scatter ...). Our formula therefore works fine for large, but not for small angles.

2. Nuclear Phenomenology

Problem 2.1 *Calculate the approximate density of nuclear matter in* gm/cm^3. *What would be the mass of a neutron star that had the diameter of an orange?*

Nuclear matter consists of tightly packed protons and neutrons. Therefore, to calculate the density of nuclear matter, it is sufficient to calculate the density of nucleons. As we know from Eq. (2.2) of the text, both the proton and the neutron have approximately the same mass (and their sizes are comparable) so that calculating the density of the proton is sufficient for our purpose. For the proton, we have

$$m_p \approx 938\,\text{MeV}/c^2 \approx 1.67 \times 10^{-24}\,\text{g}. \tag{2.1}$$

Parenthetically, we note that this leads to a relationship between the two units of energy:

$$\begin{aligned} 1\,\text{erg} &= 1\,\text{g} \times 1\,(\text{cm/sec})^2 \\ &\approx \frac{938\,\text{MeV}}{1.67 \times 10^{-24} c^2} \times 1\,(\text{cm/sec})^2 \\ &\approx \frac{938 \times 10^{24}\,\text{MeV}}{1.67 \times (3 \times 10^{10})^2\,\text{cm}^2/\text{sec}^2} \times \text{cm}^2/\text{sec}^2 \\ &\approx 6.2 \times 10^5\,\text{MeV} = 6.2 \times 10^{11}\,\text{eV}. \end{aligned} \tag{2.2}$$

Returning to our problem, the rms "charge radius" of the proton is $\approx 0.9 \times 10^{-13}$ cm. This differs somewhat from the approximate formula given for nuclear size in Eq. (2.16) of the text, which would suggest that for $A = 1$ we have

$$R_p \approx 1.2 \times 10^{-13} \text{ cm.} \tag{2.3}$$

Treating the proton as a sphere with the above radius, the density of proton can be calculated as

$$\begin{aligned}\rho_p &= \frac{m_p}{V_p} = \frac{m_p}{\frac{4}{3}\pi R_p^3} \\ &\approx \frac{1.67 \times 10^{-24}\,\text{g}}{4 \times (1.2 \times 10^{-39})^3\,\text{cm}^3} \\ &\approx 2.4 \times 10^{14}\ \text{g/cm}^3.\end{aligned} \tag{2.4}$$

This represents an approximate density of nuclear matter.

If we assume a diameter of a neutron star approximately the size of an orange:

$$d_{\text{NS}} = 2R_{\text{NS}} \approx 10\,\text{cm}, \tag{2.5}$$

then its volume will be given by

$$V_{\text{NS}} = \frac{4}{3}\pi R_{\text{NS}}^3. \tag{2.6}$$

The mass of the neutron star would therefore approximately equal:

$$\begin{aligned}M_{\text{NS}} = V_{\text{NS}}\rho_p &= \frac{4}{3}\pi R_{\text{NS}}^3 \times \frac{m_p}{\frac{4}{3}\pi R_p^3} \\ &= \left(\frac{R_{\text{NS}}}{R_p}\right)^3 m_p \approx \left(\frac{5\,\text{cm}}{1.2 \times 10^{-13}\,\text{cm}}\right)^3 \times 1.67 \times 10^{-24}\,\text{g} \\ &\approx 1.2 \times 10^{17}\ \text{g}.\end{aligned} \tag{2.7}$$

Problem 2.2 *Calculate the difference between the binding energy of a nucleus of* ^{12}C *and the sum of the binding energies of three* ^{4}He *nuclei (α-particles). Assuming that* ^{12}C *is composed of three α-particles in a triangular structure, with three effective "α-bonds" between them, what would be the binding energy per α-bond? (See CRC Handbook for Chemistry and Physics for mass values.)*

From the *CRC Handbook* we can find that

$$\begin{aligned} &\text{Binding energy of } ^{12}\text{C} \approx -92.16\,\text{MeV}, \\ &\text{Binding energy of } ^{4}\text{He} \approx -28.29\,\text{MeV}. \end{aligned} \tag{2.8}$$

Therefore, the difference in the binding energy between ^{12}C and three ^{4}He nuclei is

$$\Delta(\text{B.E.}) \approx -92.16\,\text{MeV} - 3 \times (-28.29)\,\text{MeV} = -7.29\,\text{MeV}. \tag{2.9}$$

This shows that ^{12}C is more tightly bound than three ^{4}He nuclei and is therefore stable to decay into 3 α particles.

If we assume that the ^{12}C nucleus consists of three ^{4}He nuclei in a triangular form with three α bonds, this extra binding energy must represent the binding energy of the three α bonds. Therefore, the binding energy per α bond is given by

$$\delta = \frac{\Delta(\text{B.E.})}{3} \approx -\frac{7.29\,\text{MeV}}{3} \approx -2.43\,\text{MeV}. \tag{2.10}$$

Problem 2.3 *Calculate the binding energy of the last neutron in ^{4}He and the last proton in ^{16}O. How do these compare with $\frac{B}{A}$ for these nuclei? What does this tell you about the stability of ^{4}He relative to ^{3}He, and of ^{16}O relative to ^{15}N? [Hint: the binding energy of the last neutron needed to form a nucleus (A, Z) is given by $[M(A-1, Z) + m_n - M(A, Z)]c^2$. An analogous expression holds for the last proton.]*

From the *CRC Handbook* (or the footnote on p. 34 of the text), we have

$$\begin{aligned} m_p &\approx 1.00728\,\text{amu} = 1.67 \times 10^{-24}\,\text{g} \approx 938.27\,\text{MeV}/c^2 \\ \text{or} \quad 1\,\text{amu} &\approx 1.66 \times 10^{-24}\,\text{g} \approx 931.5\,\text{MeV}/c^2. \end{aligned} \tag{2.11}$$

We are using the unified atomic mass units, sometimes also denoted by "u". In these "amu" units, the masses of the proton and neutron (given in Eq. (2.2) of the text) can be written as

$$m_p \approx 1.00728\,\text{amu}, \quad m_n \approx 1.00867\,\text{amu}. \tag{2.12}$$

To calculate the B.E. of the last nucleon, we use atomic masses rather than nuclear masses, because these are more readily available, and the masses of the electrons cancel out in the formulas. (There are very small corrections from electron binding energies in atoms that we can ignore.) The masses in question are

$$\begin{aligned} M(^1\text{H}) &\approx 1.0078\,\text{amu}, & m_n &\approx 1.0087\,\text{amu}, \\ M(^3\text{He}) &\approx 3.0160\,\text{amu}, & M(^4\text{He}) &= 4.0026\,\text{amu}, \\ M(^{15}\text{N}) &= 15.0001\,\text{amu}, & M(^{16}\text{O}) &= 15.9949\,\text{amu}. \end{aligned} \tag{2.13}$$

If we define

$$\Delta = (M(A-1, Z) + m_n - M(A, Z))c^2, \tag{2.14}$$

we can calculate the B.E. of the last neutron in ^{4}He as follows:

$$\begin{aligned} \text{B.E.} = -\Delta &= -\big(M(^3\text{He}) + m_n - M(^4\text{He})\big)c^2 \\ &\approx -(3.0160 + 1.0087 - 4.0026)\,\text{amu} \times c^2 \\ &\approx -0.0221 \times 931.5\,\text{MeV}/c^2 \times c^2 \\ &\approx -20.586\,\text{MeV}, \end{aligned} \tag{2.15}$$

where we have used (2.11).

Similarly, defining

$$\begin{aligned} \Delta &= (M(A-1, Z-1) + m_p - M(A, Z))c^2 \\ &= (M(A-1, Z-1) + M(^1\text{H}) - M(A, Z))c^2, \end{aligned} \tag{2.16}$$

where the second form is appropriate for atomic masses, we can calculate the B.E. of the last proton in ^{16}O

$$\begin{aligned} \text{B.E.} = -\Delta &= -\big(M(^{15}\text{N}) + M(^1\text{H}) - M(^{16}\text{O})\big)c^2 \\ &\approx -(15.0001 + 1.0078 - 15.9949)\,\text{amu} \times c^2 \\ &\approx -0.0130 \times 931.5\,\text{MeV}/c^2 \times c^2 \\ &\approx -12.109\,\text{MeV}. \end{aligned} \tag{2.17}$$

The average B.E. per nucleon for ^{4}He can be calculated using Eq. (2.6) of the text

$$\begin{aligned} \left.\frac{\text{B.E.}}{A}\right|_{^4\text{He}} &= -\frac{1}{4}\left(2m_p + 2m_n - M(^4\text{He})\right)c^2 \\ &= -\frac{1}{4}\big(2M(^1\text{H}) + 2m_n - M(^4\text{He})\big)c^2 \end{aligned}$$

$$
\begin{aligned}
&\approx -\frac{1}{4}(2.0156 + 2.0174 - 4.0026)\,\text{amu} \times c^2 \\
&\approx -\frac{1}{4} \times 0.0304 \times 931.5\,\text{MeV}/c^2 \times c^2 \\
&\approx -6.99\,\text{MeV}. \qquad (2.18)
\end{aligned}
$$

Similarly, for ^{16}O we have

$$
\begin{aligned}
\left.\frac{\text{B.E.}}{A}\right|_{^{16}\text{O}} &= -\frac{1}{16}\left(8m_p + 8m_n - M(^{16}\text{O})\right)c^2 \\
&= -\frac{1}{16}\left(8M(^{1}\text{H}) + 8m_n - M(^{16}\text{O})\right)c^2 \\
&\approx -\frac{1}{16}(8.0624 + 8.0696 - 15.9949)\,\text{amu} \times c^2 \\
&\approx -\frac{1}{16} \times 0.1371 \times 931.5\,\text{MeV}/c^2 \times c^2 \\
&\approx -7.99\,\text{MeV}. \qquad (2.19)
\end{aligned}
$$

Both Eqs. (2.18) and (2.19) correspond approximately to the B.E. per nucleon for low mass nuclei $\left(\frac{B}{A} = -\frac{\text{B.E.}}{A} \approx 8\,\text{MeV}\right)$. Furthermore, comparing with Eqs. (2.15) and (2.17), we conclude that the B.E. of the last neutron in ^{4}He and the last proton in ^{16}O are lower (more negative) than the average B.E. per nucleon for these nuclei. Correspondingly, we see that ^{4}He is far more stable than ^{3}He, and, likewise, ^{16}O is more stable than ^{15}N.

Problem 2.4 *Starting with cgs quantities, calculate the value of* $\mu_B = \frac{e\hbar}{2m_e c}$, *and convert it to* MeV/T *units.* (*Hint: you can relate forces and magnetic fields through the Lorentz force* $\vec{F} = \frac{q\vec{v}\times\vec{B}}{c}$.)

In CGS (cm-g-sec) units, electric charge is given in *esu* and the magnetic field in *Gauss* (G). In these units, force and energy are in *dynes* and *ergs*, respectively. In the MKS (m-kg-sec) system, the unit of electric charge is the *Coulomb* (C) and the magnetic field is given in *Tesla* (T). In these units, force and energy are in *Newtons* and *Joules*, respectively. Tesla is related to Gauss as

$$
1\,\text{T} = 10^4\,\text{G} \quad \text{or} \quad 1\,\text{G} = 10^{-4}\,\text{T}. \qquad (2.20)
$$

The magnitude of the charge of an electron (proton) is

$$
e = 4.8 \times 10^{-10}\,\text{esu} = 1.6 \times 10^{-19}\,\text{C}. \qquad (2.21)
$$

From the definition of the Lorentz force in the CGS system and the definition of work (energy)

$$\vec{F} = q\,\frac{\vec{v}}{c} \times \vec{B}, \quad E = |\vec{F}|r, \tag{2.22}$$

it follows that

$$\begin{aligned} 1\,\text{erg} &= 1\,\text{esu-G-cm}, \\ \text{or} \quad 1\,\text{esu-cm} &= 1\,\text{erg/G}. \end{aligned} \tag{2.23}$$

We note from the definition of the Bohr magneton that

$$\begin{aligned} \mu_{\text{B}} &= \frac{e\hbar}{2m_e c} = e \times \frac{\hbar c}{2m_e c^2} \\ &\approx 4.8 \times 10^{-10}\,\text{esu} \times \frac{197\,\text{MeV} - \text{F}}{2 \times 0.511\,\text{MeV}} \\ &= \frac{4.8 \times 197}{1.022} \times 10^{-10} \times 10^{-13}\,\text{esu-cm} \\ &= \frac{4.8 \times 197}{1.022} \times 10^{-23}\,\text{erg/G} \\ &\approx \frac{4 \times 197}{1.022} \times 10^{-23} \times \frac{6.2 \times 10^{5}\,\text{MeV}}{\text{G}} \\ &\approx 5.8 \times 10^{-15}\,\text{MeV/G} \\ &= 5.8 \times 10^{-15}\frac{\text{MeV}}{10^{-4}\,\text{T}} = 5.8 \times 10^{-11}\,\text{MeV/T}, \end{aligned} \tag{2.24}$$

which agrees approximately with the value given in Eq. (2.18) of the text. Here, in the intermediate steps we have used (2.2), which converts "erg" to "MeV".

Problem 2.5 *Assume that the spin of a proton can be represented by a positive pion moving at a speed c in a circular orbit of radius* 10^{-13} cm *about a neutral center. Calculate the current and the magnetic moment associated with this motion. Compare this with the known magnetic moment of the proton. (Hint: recall that using cgs units you can write a magnetic moment* $\vec{\mu} = (\frac{I}{c})\vec{A}$, *where I is the current flowing around the area A.)*

For this simple model of proton motion, we have

$$\begin{aligned} e &\approx 4.8 \times 10^{-10}\,\text{esu}, \\ r &= \text{radius of circular orbit} = 10^{-13}\,\text{cm}, \\ c &= \text{speed of rotation} = 3 \times 10^{10}\,\text{cm/sec}. \end{aligned} \tag{2.25}$$

It follows that the period of rotation is given by

$$T = \frac{2\pi r}{c} \approx \frac{6 \times 10^{-13}\,\text{cm}}{3 \times 10^{10}\,\text{cm/sec}} = 2 \times 10^{-23}\,\text{sec}. \tag{2.26}$$

The current (rate of change of charge with time) associated with the motion of the charge is given by

$$I = \frac{e}{T} \approx \frac{4.8 \times 10^{-10}\,\text{esu}}{2 \times 10^{-23}\,\text{sec}} = 2.4 \times 10^{13}\,\text{esu/sec}. \tag{2.27}$$

As we know from classical electromagnetism, in CGS units, the magnetic moment associated with a circular loop carrying current I is given by

$$\mu = \frac{I}{c} \times A, \tag{2.28}$$

where A represents the area enclosed by the current loop. Applying this to the simple model of proton motion, we obtain

$$\begin{aligned} \mu_p &= \frac{I}{c} \times \pi r^2 \approx \frac{2.4 \times 10^{13}\,\text{esu/sec}}{3 \times 10^{10}\,\text{cm/sec}} \times 3 \times 10^{-26}\,\text{cm}^2 \\ &= 2.4 \times 10^{-23}\,\text{esu-cm} = 2.4 \times 10^{-23}\,\text{erg/G} \\ &\approx 2.4 \times 10^{-23} \times \frac{6.2 \times 10^{5}\,\text{MeV}}{10^{-4}\,\text{T}} \\ &\approx 1.489 \times 10^{-13}\,\text{MeV/T}, \end{aligned} \tag{2.29}$$

where we have used (2.2). The result for the magnetic moment can be compared with the measured value of the magnetic moment of the proton (given in Eq. (2.20) of the text)

$$\begin{aligned}\mu_p &\approx 2.79\mu_N = 2.79 \times \frac{m_e}{m_N} \times \mu_B \\ &\approx 2.79 \times \frac{0.511\,\text{MeV}/c^2}{938.27\,\text{MeV}/c^2} \times 5.79 \times 10^{-11}\,\text{MeV/T} \\ &\approx 2.79 \times 3.15 \times 10^{-14}\,\text{MeV/T} \approx 8.8 \times 10^{-14}\,\text{MeV/T}. \end{aligned} \tag{2.30}$$

Hence, this simple model leads to good agreement for the order of magnitude of the magnetic moment of the proton.

Problem 2.6 *We argued previously that the π^+ mesons in Fig. 2.2 scattered not from individual nucleons, but rather (coherently) from the entire nuclei. In fact, the first minima ($n = 1$) corresponded to $\theta \approx \frac{nh}{2Rp}$, with R being consistent with $1.2\,A^{\frac{1}{3}}$. At higher energies, when larger momenta can be transferred to nuclei, it is possible to dislodge a single proton or neutron from the nucleus. When this happens, the π^+ mesons can be termed to scatter elastically from quasi "free" nucleons. How would this affect the diffraction pattern in Fig. 2.2? What about if you could scatter from very small point-like constituents within nucleons? (Would the fact that a π^+ is not a point particle affect your answer?)*

From the classical formula for diffraction from a sphere of radius R, we know that the minima of the diffraction pattern occur at

$$2R\sin\theta_{\text{min}}^{(n)} = n\lambda, \quad n = 1, 2, \ldots, \tag{2.31}$$

where λ denotes the wavelength of the incident beam. For small angles, this leads to

$$\theta_{\text{min}}^{(n)} = \frac{n\lambda}{2R} = \frac{nh}{2Rp}, \tag{2.32}$$

where we have used de Broglie's hypothesis to relate the momentum of a particle to its wavelength

$$\lambda = \frac{h}{p}. \tag{2.33}$$

Here $h = 2\pi\hbar$ is Planck's constant and p denotes the magnitude of the momentum of the incident particle. The first minimum occurs at

$$\theta_{\text{min}}^{(1)} = \frac{h}{2Rp}. \tag{2.34}$$

If the incident π^+ has low energy, it will primarily scatter from the target nuclei. The first minimum of the diffraction pattern will therefore occur at an angle $\theta_{\text{min}}^{(1)}$ corresponding to

$$R = 1.2 \times 10^{-13} A^{\frac{1}{3}} \text{ cm}, \tag{2.35}$$

where A denotes the mass number of the target nucleus. As the momentum (and therefore energy) of the π^+ increases, it can penetrate deeper into the nucleus and scatter from individual nucleons. In the limit that it has infinite momentum, it can scatter from effectively "free" nucleons within the nucleus, in which case the first minimum of the diffraction pattern will occur at values corresponding to

$$R = 1.2 \times 10^{-13} \text{ cm}. \tag{2.36}$$

This angle will be relatively larger than the minimum angle for scattering from the nucleus since the size of the nucleon is smaller than the size of the nucleus. At intermediate values of momentum, both components of diffraction will be present, and hence these two contributions to the scattering must be added together. The "coherent" scattering from the entire nucleus arises from a sum of amplitudes for scattering off individual nucleons close to $\theta = 0$, with the final rate given by the square of the sum of the amplitudes. While the "incoherent" scattering from individual "free" nucleons arises from a sum over individual intensities. Thus, naively, the rate for coherent scattering is expected to go as $|A|^2$ and for incoherent scattering as $|A|$.

If the scattering is from even smaller constituents within nucleons, then an even more diffuse component (larger θ values) must be added to the spectrum. If pions are not point-like, then $R \approx \sqrt{R_\pi^2 + R_{\text{other}}^2}$.

Problem 2.7 *Normally, in optics, one looks at the diffraction pattern as a function of angle θ. In this case, the value of θ at the first minimum changes with wavelength or momentum. Can you see any advantage to using a variable such as $q^2 \approx p_T^2 \approx (p\theta)^2$ to examine diffraction patterns at different scattering energies? Sketch how the pattern might look for scattering of π^+ mesons of different energies from nuclear targets. Now, as energy increases, and larger q^2*

become possible, what would be the effect of having nucleon substructure within the nucleus? What about point substructure within the nucleon? (Does your answer depend on whether the π^+ has such substructure?)

As is clear from Eq. (2.32), the positions of the minima of the diffraction pattern depend on the momentum of the incident particle. On the other hand, for small angles, if we identify the momentum transfer as

$$q^2 = p^2 \sin^2\theta \approx (p\theta)^2, \tag{2.37}$$

then using (2.32) we can write the locations of the minima of the diffraction pattern as

$$\left(q_{\min}^{(n)}\right)^2 \approx \left(p\theta_{\min}^{(n)}\right)^2 = \frac{n^2h^2}{4R^2}. \tag{2.38}$$

It is clear that in this q^2 variable, the locations of the minima of the diffraction pattern depend only on R, and are independent of the incident momentum. (As it turns out, the size of the nucleon appears to grow with incident energy, and the diffraction pattern shrinks as the scattering energy increases. This growth of the nucleon cross section is not completely understood.)

Problem 2.8 *What are the frequencies that correspond to typical splitting of lines for nuclear magnetic moments in magnetic fields of* ≈ 5 tesla?

The interaction of a particle of magnetic moment $\vec{\mu}$ with a magnetic field $\vec{B}$ leads to a typical shift in the energy

$$\Delta E = \vec{\mu} \cdot \vec{B} \approx \mu B, \tag{2.39}$$

where μ, B denote the magnitudes of the magnetic moment and the magnetic field, respectively. For particles with typical nuclear magnetic moments in the presence of a 5 T magnetic field, we have:

$$\begin{aligned} \Delta E = \mu_N B &\approx 3.15 \times 10^{-14}\,\text{MeV/T} \times 5\,\text{T} \\ &\approx 1.57 \times 10^{-13}\,\text{MeV}, \end{aligned} \tag{2.40}$$

where we have used (2.30). This shift in the energy will be reflected in a shift in the frequency of the lines by

$$\begin{aligned}\Delta\nu &= \frac{\Delta E}{h} = \frac{\Delta E \times c}{2\pi\hbar c} \\ &\approx \frac{1.57 \times 10^{-13}\,\text{MeV} \times 3 \times 10^{10}\,\text{cm/sec}}{6 \times 197\,\text{MeV} - \text{F}} \\ &\approx 3.95 \times 10^{7}/\text{sec} \approx 39\,\text{MHz}.\end{aligned} \tag{2.41}$$

The corresponding wavelength is given by

$$\lambda = \frac{c}{\nu} \approx \frac{3 \times 10^{10}\,\text{cm/sec}}{3.95 \times 10^{7}/\text{sec}} \approx 7.5 \times 10^{2}\,\text{cm} = 7.5\,\text{m}. \tag{2.42}$$

This is in the upper range of short-wave radio frequencies (RF).

Problem 2.9 *Show that when non-relativistic neutrons of kinetic energy E_0 collide head-on with stationary nuclei of mass number A, the smallest energy that elastically-scattered neutrons can have is given approximately by*

$$E_{\text{min}} = E_0 \left(\frac{A-1}{A+1}\right)^2 .$$

What will be the approximate energies of the neutrons after one, two, and any number j of such consecutive collisions, if the target nucleus is hydrogen, carbon, and iron?

For a neutron incident on a much heavier target nucleus, we see from Eqs. (1.1) and (1.2) of the text that

$$\begin{aligned} m_n v_n^2 &= m_n v_0^2 - m_t v_t^2 \\ &= m_n v_0^2 - m_t \times \frac{m_n^2}{m_t^2} \times (\vec{v}_0 - \vec{v}_n)^2 \\ &= m_n \left[\left(1 - \frac{m_n}{m_t}\right) v_0^2 - \frac{m_n}{m_t}\, v_n^2 + \frac{2m_n}{m_t}\, v_0 v_n \cos\theta\right] \end{aligned}$$

$$\text{or} \quad (m_t + m_n) v_n^2 = (m_t - m_n) v_0^2 + 2 m_n v_0 v_n \cos\theta, \tag{2.43}$$

where θ is the scattering angle and v_0, v_n represent the magnitudes of the incident and scattered neutron velocities. It is clear from (2.43) that, for $m_t > m_n$, both v_n^2 and thereby the kinetic energy of the scattered neutron will be minimum when $\theta = \pi$, that is, when the neutron scatters backwards.

For $\theta = \pi$, we note that Eq. (2.43) takes the form

$$(m_t + m_n)v_n^2 + 2m_n v_n v_0 - (m_t - m_n)v_0^2 = 0, \tag{2.44}$$

which can be solved to yield

$$v_n = \frac{-2m_n v_0 \pm \sqrt{4m_n^2 v_0^2 + 4(m_t^2 - m_n^2)v_0^2}}{2(m_t + m_n)} = \frac{-m_n \pm m_t}{m_t + m_n}\, v_0. \tag{2.45}$$

Since $v_n > 0$, it follows that when $m_t > m_n$,

$$v_n = \frac{m_t - m_n}{m_t + m_n}\, v_0. \tag{2.46}$$

Consequently, the energy of the neutron when scattered backwards is given by

$$E_n = \frac{1}{2}m_n v_n^2 = \left(\frac{m_t - m_n}{m_t + m_n}\right)^2 \frac{1}{2}m_n v_0^2 = \left(\frac{m_t - m_n}{m_t + m_n}\right)^2 E_0. \tag{2.47}$$

Since the neutron mass corresponds to a mass number $A \approx 1$, if the target nucleus has a mass number A, we can write

$$E_n = \left(\frac{A-1}{A+1}\right)^2 E_0. \tag{2.48}$$

Thus, we see that everytime a neutron is scattered backwards, its energy will be reduced by a factor of

$$\left(\frac{A-1}{A+1}\right)^2. \tag{2.49}$$

Consequently, if it scatters backwards j consecutive times, its energy will be given by

$$E_{n,j} = \left(\frac{A-1}{A+1}\right)^{2j} E_0, \quad j = 1, 2, 3, \ldots. \tag{2.50}$$

For scattering from hydrogen, carbon and iron ($A = 1, 12, 56$, respectively), we can tabulate the energy of the neutron after successive backward scatterings as

Collision #	^{1}H	^{12}C	^{56}Fe
1	≈ 0	$\left(\frac{11}{13}\right)^2 E_0$	$\left(\frac{55}{57}\right)^2 E_0$
2	≈ 0	$\left(\frac{11}{13}\right)^4 E_0$	$\left(\frac{55}{57}\right)^4 E_0$
3	≈ 0	$\left(\frac{11}{13}\right)^6 E_0$	$\left(\frac{55}{57}\right)^6 E_0$
$\vdots$	$\vdots$	$\vdots$	$\vdots$
j	≈ 0	$\left(\frac{11}{13}\right)^{2j} E_0$	$\left(\frac{55}{57}\right)^{2j} E_0$

Problem 2.10 *Using the results of Problem 2.9, calculate the number of collisions needed to reduce the energy of a* 2 MeV *neutron to* 0.1 MeV *through elastic collisions between the neutron and carbon nuclei.*

As we have seen in Eq. (2.50) in the previous problem, after j successive backward scatterings, the energy of the scattered neutron is given by

$$E_{n,j} = \left(\frac{A-1}{A+1}\right)^{2j} E_0. \tag{2.51}$$

If the scattering is from ^{12}C nuclei, we have

$$E_{n,j} = \left(\frac{12-1}{12+1}\right)^{2j} E_0 = \left(\frac{11}{13}\right)^{2j} E_0. \tag{2.52}$$

If $E_0 = 2\,\text{MeV}$ and $E_{n,j} = 0.1\,\text{MeV}$, we obtain

$$\begin{aligned} 0.1\,\text{MeV} &= \left(\frac{11}{13}\right)^{2j} \times 2\,\text{MeV} \\ \text{or} \quad \left(\frac{11}{13}\right)^{2j} &= \frac{0.1\,\text{MeV}}{2\,\text{MeV}} = 0.05 \\ \text{or} \quad 2j \ln \frac{11}{13} &= \ln(0.05) \\ \text{and} \quad j \approx \frac{\ln(0.05)}{2 \times \ln(0.85)} &\approx \frac{(-3)}{2 \times 0.165} \approx 9.1. \end{aligned} \tag{2.53}$$

Therefore, it would take about 9 consecutive backwards scattering of the neutron to reduce its energy from 2 MeV down to 0.1 MeV.

Problem 2.11 *For $q^2 \ll 1$, the exponential in the elastic form factor of Eq. (2.13) can be approximated as $1 + i\vec{k}\cdot\vec{r} - \frac{1}{2}(\vec{k}\cdot\vec{r})^2$, where $\vec{k} = \frac{1}{\hbar}\vec{q}$. Calculate $|F(q)|^2$ in terms of a root-mean-square radius of the charge distribution $R = \sqrt{\langle r^2\rangle}$, for $\rho(r)$ described by (a) a uniform distribution of charge within $r = R$, and (b) a Gaussian form $\rho(r) \approx e^{-\frac{2r^2}{R^2}}$, and show that in both cases $|F(q)|^2$ falls off approximately exponentially with q^2. (Hint: Use symmetry arguments to eliminate the $\vec{k}\cdot\vec{r}$ term by recognizing that $\vec{k}\cdot\vec{r} = k_x x + k_y y + k_z z$. Also, note that for a spherically symmetric $\rho(r)$, $\langle x^2\rangle = \langle y^2\rangle = \langle z^2\rangle = \frac{1}{3}\langle r^2\rangle$, and $\langle r^2\rangle = \int 4\pi r^2 dr\, r^2\rho(r)$.)*

In Eq. (2.13) of the text, the form factor is given by

$$F(\vec{q}) = \int_{\text{all space}} \mathrm{d}^3 r\ \rho(\vec{r}) e^{\frac{i}{\hbar}\vec{q}\cdot\vec{r}}. \tag{2.54}$$

In spherical coordinates, the exponent can be written as

$$\frac{i}{\hbar}\vec{q}\cdot\vec{r} = \frac{i}{\hbar} qr\cos\theta, \tag{2.55}$$

where q, r represent the magnitudes of the three-vectors $\vec{q}, \vec{r}$, respectively. For $q^2 \ll 1$, we can Taylor expand the exponential to write

$$F(\vec{q}) \approx \int_{\text{all space}} \mathrm{d}^3 r\ \rho(\vec{r}) \left(1 + \frac{i}{\hbar} qr\cos\theta - \frac{1}{2\hbar^2} q^2 r^2 \cos^2\theta\right). \tag{2.56}$$

Furthermore, if the charge distribution is spherically symmetric, namely,

$$\rho(\vec{r}) = \rho(r), \tag{2.57}$$

we can simplify Eq. (2.56) by integrating over angular coordinates, which yields

$$\begin{aligned} F(\vec{q}) &\approx \int r^2 \mathrm{d}r\, \mathrm{d}(\cos\theta) \mathrm{d}\phi\ \rho(r) \left(1 + \frac{i}{\hbar} qr\cos\theta - \frac{1}{2\hbar^2} q^2 r^2 \cos^2\theta\right) \\ &= 4\pi \int \mathrm{d}r\, r^2 \rho(r) \left(1 - \frac{q^2 r^2}{6\hbar^2}\right). \end{aligned} \tag{2.58}$$

If the charge distribution is normalized to unity

$$\int_{\text{all}} \mathrm{d}^3 r\, \rho(r) = 4\pi \int \mathrm{d}r\, r^2 \rho(r) = 1, \tag{2.59}$$

and if we identify the mean square radius of the charge distribution as

$$r_{\rm rms}^2 = \langle r^2 \rangle = \int_{\rm all\ space} {\rm d}^3 r\, \rho(r) \times r^2 = 4\pi \int {\rm d}r\, r^4 \rho(r). \qquad (2.60)$$

We can rewrite the form factor in Eq. (2.58) as

$$\begin{aligned} F(\vec{q}) &\approx 1 - \frac{q^2}{6\hbar^2} \langle r^2 \rangle \\ &\approx e^{-\frac{q^2}{6\hbar^2} \langle r^2 \rangle}, \end{aligned} \qquad (2.61)$$

where we have used the fact that $q^2 \ll 1$ to write the form factor in the approximate exponentiated form. It follows now that

$$|F(\vec{q})|^2 \approx e^{-\frac{q^2}{3\hbar^2} \langle r^2 \rangle}. \qquad (2.62)$$

This shows that the form factor falls off exponentially with q^2.

Let us next consider two special cases. First, when there is a uniform charge distribution within a spherical volume of radius R. Here we have

$$\rho(r) = \begin{cases} \dfrac{1}{\frac{4}{3}\pi R^3} & r \leq R, \\ 0 & r > R. \end{cases} \qquad (2.63)$$

Clearly, the charge distribution is normalized to unity, and we have

$$\begin{aligned} \langle r^2 \rangle_{\rm uniform} &= 4\pi \int {\rm d}r\, r^4 \rho(r) 4\pi \times \frac{1}{\frac{4}{3}\pi R^3} \int_0^R {\rm d}r\, r^4 \\ &= \frac{3}{R^3} \times \frac{R^5}{5} = \frac{3R^2}{5}, \end{aligned} \qquad (2.64)$$

so that from (2.62) we get

$$|F_{\rm uniform}(\vec{q})|^2 \approx e^{-\frac{q^2 R^2}{5\hbar^2}}. \qquad (2.65)$$

The second example corresponds to a Gaussian charge distribution of the form

$$\rho(r) = \left(\frac{2}{\pi R^2}\right)^{\frac{3}{2}} e^{-\frac{2r^2}{R^2}}, \qquad (2.66)$$

where R is some fixed length scale that reflects the size of the distribution. We can check the normalization of this distribution by noting that

$$\begin{aligned}\int_{\text{all space}} \mathrm{d}^3r\,\rho(r) &= 4\pi \int \mathrm{d}r\, r^2 \rho(r) \\ &= 4\pi \times \left(\frac{2}{\pi R^2}\right)^{\frac{3}{2}} \int_0^\infty \mathrm{d}r\, r^2 e^{-\frac{2r^2}{R^2}} \\ &= 4\pi \times \left(\frac{2}{\pi R^2}\right)^{\frac{3}{2}} \left(\frac{R}{\sqrt{2}}\right)^3 \int_0^\infty \mathrm{d}x\, x^2 e^{-x^2} \\ &= 4\pi \times \left(\frac{2}{\pi R^2}\right)^{\frac{3}{2}} \left(\frac{R^2}{2}\right)^{\frac{3}{2}} \frac{1}{2}\Gamma\left(\frac{3}{2}\right) \\ &= 4\pi \times \left(\frac{2}{\pi R^2}\right)^{\frac{3}{2}} \left(\frac{R^2}{2}\right)^{\frac{3}{2}} \frac{1}{2} \times \frac{\sqrt{\pi}}{2} = 1. \end{aligned} \tag{2.67}$$

(Note that the distribution given in the statement of the problem is not normalized to unity.) In the intermediate steps above we defined

$$x = \frac{\sqrt{2}r}{R}, \tag{2.68}$$

and we used the "standard" definition of the "Gamma (Γ) function"

$$\Gamma(n) = 2\int_0^\infty \mathrm{d}x\, x^{2n-1} e^{-x^2}. \tag{2.69}$$

Using this distribution, we can now calculate

$$\begin{aligned}\langle r^2 \rangle_{\text{Gaussian}} &= 4\pi \int \mathrm{d}r\, r^4 \rho(r) \\ &= 4\pi \times \left(\frac{2}{\pi R^2}\right)^{\frac{3}{2}} \int_0^\infty \mathrm{d}r\, r^4 e^{-\frac{2r^2}{R^2}} \\ &= 4\pi \times \left(\frac{2}{\pi R^2}\right)^{\frac{3}{2}} \times \left(\frac{R^2}{2}\right)^{\frac{5}{2}} \int_0^\infty \mathrm{d}x\, x^4 e^{-x^2} \\ &= 4\pi \times \left(\frac{2}{\pi R^2}\right)^{\frac{3}{2}} \times \left(\frac{R^2}{2}\right)^{\frac{5}{2}} \times \frac{1}{2}\Gamma\left(\frac{5}{2}\right) \\ &= 4\pi \times \left(\frac{2}{\pi R^2}\right)^{\frac{3}{2}} \times \left(\frac{R^2}{2}\right)^{\frac{5}{2}} \times \frac{1}{2} \times \frac{3\sqrt{\pi}}{4} \\ &= \frac{3R^2}{4}, \end{aligned} \tag{2.70}$$

so that Eq. (2.62) leads to

$$|F_{\text{Gaussian}}(\vec{q})|^2 \approx e^{-\frac{q^2 R^2}{4\hbar^2}}. \tag{2.71}$$

In both cases (and, in fact, in general for spherically symmetric distributions of charge) the form factor falls off exponentially with q^2. (Needless to say, you do not have to know about Γ functions to do this problems — you can just look up these integrals in your favorite tables of definite integrals.)

3. Nuclear Models

Problem 3.1 *The Bethe–Weizsäcker formula of Eq.* (3.5) *provides an excellent representation of the mass systematics of nuclei. Show explicitly that, for fixed* A, $M(A, Z)$ *has a minimum value. Is there evidence for the "valley of stability" observed in Fig.* 2.3? *What is the stablest nucleus with* $A = 16$? *What about* $A = 208$? (*You can differentiate Eq.* (3.5), *or simply plot* M *as a function of* Z.)

From Eq. (3.5) of the text, the mass of the nucleus as a function of its charge (Z) and the nucleon number (A) is given by

$$M(A,Z) = (A-Z)m_n + Zm_p - \frac{a_1}{c^2}A + \frac{a_2}{c^2}A^{\frac{2}{3}} + \frac{a_3}{c^2}\frac{Z^2}{A^{\frac{1}{3}}} + \frac{a_4}{c^2}\frac{(A-2Z)^2}{A} \pm a_5 A^{-\frac{3}{4}}, \tag{3.1}$$

where the values of the positive coefficients $a_1, \ldots, a_5$ are given in Eq. (3.4) of the text. For a fixed value of A, the nuclear mass becomes a function of its charge alone. The value of the nuclear charge for which the mass becomes a minimum can be easily determined from

$$\left.\frac{\partial M(A,Z)}{\partial Z}\right|_{\text{fixed } A} = -m_n + m_p + \frac{2a_3 Z}{c^2 A^{\frac{1}{3}}} - \frac{4a_4(A-2Z)}{c^2 A} = 0$$

$$\text{or} \quad \frac{2Z}{c^2 A}\left(a_3 A^{\frac{2}{3}} + 4a_4\right) = \frac{1}{c^2}\left(4a_4 + (m_n - m_p)c^2\right) \tag{3.2}$$

$$\text{or} \quad Z_{\min}(A) = \frac{A}{2} \times \frac{4a_4 + (m_n - m_p)c^2}{4a_4 + a_3 A^{\frac{2}{3}}}.$$

To determine whether this stationary point is a minimum or a maximum, we note that

$$\left.\frac{\partial^2 M(A,Z)}{\partial^2 Z}\right|_{\text{fixed } A} = \frac{2}{c^2 A}\left(4a_4 + a_3 A^{\frac{2}{3}}\right) > 0. \tag{3.3}$$

We therefore conclude that the extremum in Eq. (3.2) is a minimum.

The solution for the minimum in Eq. (3.2) represents a valley of minima, similar to what is observed in Fig. 2.3 of the text. Noting from Eqs. (2.2) and (3.4) of the text that

$$a_3 = 0.72\,\text{MeV}, \quad a_4 = 23.3\,\text{MeV}, \quad (m_n - m_p)c^2 \approx 1.29\,\text{MeV}, \tag{3.4}$$

we conclude that for low values of A, the minimum in the mass occurs for

$$Z_{\text{min}} \approx \frac{A}{2}, \tag{3.5}$$

which represents the stability line. However, as A increases, the denominator becomes larger than the numerator and we have

$$Z_{\text{min}} < \frac{A}{2} \quad \text{or} \quad N > Z_{\text{min}}, \tag{3.6}$$

where we have used the fact that $A = Z + N$. This deviation from stability starts when

$$a_3 A^{\frac{2}{3}} > (m_n - m_p)c^2$$

$$\text{or} \quad A > \left(\frac{(m_n - m_p)c^2}{a_3}\right)^{\frac{3}{2}} \approx \left(\frac{1.29\,\text{MeV}}{0.72\,\text{MeV}}\right)^{\frac{3}{2}} \approx 2.2, \tag{3.7}$$

which is somewhat low compared to the experimental observations sketched in Fig. 2.3 of the text.

From the formula in (3.2) we can calculate

$$\begin{aligned} Z_{\text{min}}(A = 16) &= \frac{A}{2} \times \frac{4a_4 + (m_n - m_p)c^2}{4a_4 + 0.72A^{\frac{2}{3}}} \\ &= \frac{16}{2} \times \frac{4 \times 23.3\,\text{MeV} + 1.29\,\text{MeV}}{4 \times 23.3\,\text{MeV} + 0.72\,\text{MeV} \times (16)^{\frac{2}{3}}} \\ &\approx 8 \times 0.95 = 7.6, \end{aligned} \tag{3.8}$$

$$\begin{aligned} Z_{\text{min}}(A = 208) &= \frac{208}{2} \times \frac{4 \times 23.3\,\text{MeV} + 1.29\,\text{MeV}}{4 \times 23.3\,\text{MeV} + 0.72\,\text{MeV} \times (208)^{\frac{2}{3}}} \\ &\approx 104 \times 0.79 \approx 82.16. \end{aligned}$$

We note that for $A = 16$ and $A = 208$, we expect $^{16}O^8$ and $^{208}Pb^{82}$, respectively, to represent the most stable nuclei, and so this model leads to reasonably accurate predictions.

This result can also be obtained graphically as follows. First, we note that, for a fixed value of A, we can write the mass formula in Eq. (3.1) as

$$\begin{aligned} M(A,Z)c^2 &= \frac{4a_4 + a_3 A^{\frac{2}{3}}}{A} Z^2 - (4a_4 + (m_n - m_p)c^2)Z + C \\ &= \frac{4a_4 + a_3 A^{\frac{2}{3}}}{A} \left(Z - \frac{A}{2} \times \frac{4a_4 + (m_n - m_p)c^2}{4a_4 + a_3 A^{\frac{2}{3}}} \right)^2 \\ &\quad + \left(C - \frac{A}{4} \times \frac{(4a_4 + (m_n - m_p)c^2)^2}{4a_4 + a_3 A^{\frac{2}{3}}} \right) \\ &= \frac{4a_4 + a_3 A^{\frac{2}{3}}}{A} (Z - Z_{\min})^2 \\ &\quad + \left(C - \frac{A}{4} \times \frac{(4a_4 + (m_n - m_p)c^2)^2}{4a_4 + a_3 A^{\frac{2}{3}}} \right), \end{aligned} \tag{3.9}$$

where we have collected all the constant terms into C, namely,

$$C = Am_n c^2 - a_1 A + a_2 A^{\frac{2}{3}} + a_4 A \pm a_5 A^{-\frac{3}{4}}. \tag{3.10}$$

The mass formula depends quadratically on Z, and, when plotted against Z for fixed A, shows a minimum at $Z = Z_{\min}$ as given in (3.2). For $A = 16$ and $A = 208$, the total number of nucleons is a multiple of 4. Consequently, we can have only an even–even structure (although an odd–odd nucleus has even A, it will not correspond to a multiple of 4), so that we need to consider only the negative sign in the last term of the mass formula in (3.1). For completeness, we plot in Figs. 3.1 and 3.2 the graphs for $A = 16$ and 208, respectively, for both odd–odd and even–even nuclei.

Problem 3.2 *Using Eq.* (3.3) *compute the total binding energy and the value of* $\frac{B}{A}$ *for* $^8Be^4$, $^{12}C^6$, $^{56}Fe^{26}$ *and* $^{208}Pb^{82}$. *How do these values compare with experiment? (See CRC Handbook of Chemistry and Physics for data.)*

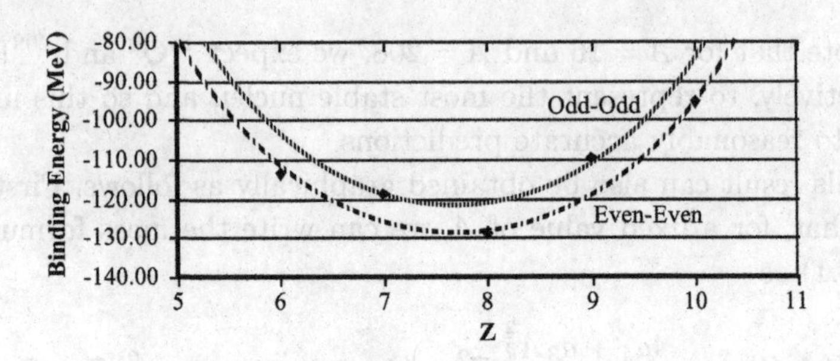

Fig. 3.1. Mass formula for fixed $A = 16$ vs. Z.

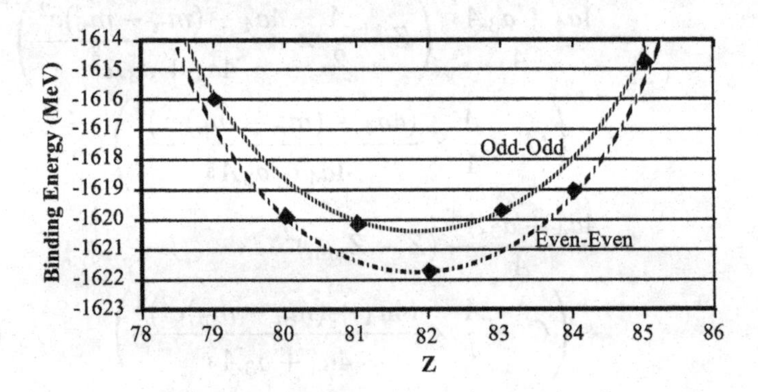

Fig. 3.2. Mass formula for fixed $A = 208$ vs. Z.

The empirical formula for the B.E. is given in Eq. (3.3) of the text

$$\begin{aligned}\text{B.E.} &= -a_1 A + a_2 A^{\frac{2}{3}} + a_3 Z^2 A^{-\frac{1}{3}} + a_4 (N-Z)^2 A^{-1} \pm a_5 A^{-\frac{3}{4}} \\ &= -a_1 A + a_2 A^{\frac{2}{3}} + a_3 Z^2 A^{-\frac{1}{3}} + a_4 (A-2Z)^2 A^{-1} \pm a_5 A^{-\frac{3}{4}},\end{aligned} \tag{3.11}$$

where the upper sign in the last term is for odd–odd nuclei while the lower sign is for the even–even nuclei. The coefficients in the expression have values given in Eq. (3.4) of the text, namely,

$$\begin{aligned}&a_1 \approx 15.6\,\text{MeV}, \quad a_2 \approx 16.8\ \text{MeV}, \quad a_3 \approx 0.72\,\text{MeV}, \\ &a_4 \approx 23.3\,\text{MeV}, \quad a_5 \approx 34\ \text{MeV}.\end{aligned} \tag{3.12}$$

The B.E. per nucleon can be obtained from this to correspond to

$$\frac{\text{B.E.}}{A} = -a_1 + a_2 A^{-\frac{1}{3}} + a_3 Z^2 A^{-\frac{4}{3}} + a_4 (A-2Z)^2 A^{-2} \pm a_5 A^{-\frac{7}{4}}. \tag{3.13}$$

We note that $^8\text{Be}^4$, $^{12}\text{C}^6$, $^{56}\text{Fe}^{26}$ and $^{208}\text{Pb}^{82}$ are all even–even nuclei. Therefore, we consider only the lower sign in the last term in Eqs. (3.11) and (3.13) in evaluating the B.E. and the B.E. per nucleon. Putting in the numbers, we obtain the table:

Table 3.1.

A	Z	B.E.$_{\text{calculated}}$ (MeV)	B.E.$_{\text{tables}}$ (MeV)	$\frac{B}{A} = -\frac{\text{B.E.}}{A}$ (MeV)
8	4	−58.99	−56.50	7.37
12	6	−93.09	−92.20	7.76
56	26	−495.48	−492.30	8.85
208	82	−1621.68	−1635.80	7.80

Problem 3.3 *You might conclude from Problem 3.2 that $^8Be^4$ is stable. This is, in fact, not the case. Can you provide a model to explain this result? (Hint: see Problem 2.2.)*

As we have already seen in Problem 2.2,

$$\text{B.E. of } ^4\text{He}^2 = -28.29\,\text{MeV}. \tag{3.14}$$

Therefore, we see from Table 3.1 that the difference in the B.E. of $^8\text{Be}^4$ and that of two $^4\text{He}^2$ nuclei is given by

$$\begin{aligned}\Delta &= -56.50\,\text{MeV} - 2\times(-28.29)\,\text{MeV}\\ &= (-56.50 + 56.58)\,\text{MeV} = 0.08\,\text{MeV}.\end{aligned} \tag{3.15}$$

As a result, we see that $^8\text{Be}^4$ can decay into two $^4\text{He}^2$ nuclei, thereby releasing 80 keV of energy:

$$^8\text{Be}^4 \rightarrow {}^4\text{He}^2 + {}^4\text{He}^2 + 80\,\text{keV}. \tag{3.16}$$

Problem 3.4 *Calculate the binding energy of the last neutron in $^{15}N^7$ and of the last proton in $^{15}O^8$, and contrast with the last neutron in $^{16}N^7$ and in $^{16}O^8$.*

From the *CRC Handbook*, we know that

$$\begin{aligned}
M(^1\mathrm{H}^1) &= 1.0078\,\mathrm{amu}, & m_n &= 1.0087\,\mathrm{amu}, \\
M(^{14}\mathrm{N}^7) &= 14.0031\,\mathrm{amu}, & M(^{15}\mathrm{N}^7) &= 15.0001\,\mathrm{amu}, \\
M(^{16}\mathrm{N}^7) &= 16.0061\,\mathrm{amu}, & M(^{15}\mathrm{O}^8) &= 15.0030\,\mathrm{amu}, \\
M(^{16}\mathrm{O}^8) &= 15.9949\,\mathrm{amu}.
\end{aligned} \tag{3.17}$$

Using these values and the conversion between "amu" and "MeV" units, we can calculate the binding energy of the last neutron in $^{15}\mathrm{N}^7$

$$\begin{aligned}
\mathrm{B.E.} &= -(M(^{14}\mathrm{N}^7) + m_n - M(^{15}\mathrm{N}^7))c^2 \\
&= -(14.0031 + 1.0087 - 15.0001)\,\mathrm{amu} \times c^2 \\
&\approx -0.0117 \times 931.5\,\mathrm{MeV}/c^2 \times c^2 = -10.8985\,\mathrm{MeV}.
\end{aligned} \tag{3.18}$$

Similarly, the binding energy of the last proton in $^{15}\mathrm{O}^8$ is

$$\begin{aligned}
\mathrm{B.E.} &= -(M(^{14}\mathrm{N}^7) + M(^1\mathrm{H}^1) - M(^{15}\mathrm{O}^8))c^2 \\
&= -(14.0031 + 1.0078 - 15.0030)\,\mathrm{amu} \times c^2 \\
&\approx -0.0079 \times 931.5\,\mathrm{MeV}/c^2 \times c^2 = -7.3588\,\mathrm{MeV}.
\end{aligned} \tag{3.19}$$

Furthermore, the binding energy for the last neutron in $^{16}\mathrm{N}^7$ is given by

$$\begin{aligned}
\mathrm{B.E.} &= -(M(^{15}\mathrm{N}^7) + m_n - M(^{16}\mathrm{N}^7))c^2 \\
&= -(15.0001 + 1.0087 - 16.0061)\,\mathrm{amu} \times c^2 \\
&\approx -0.0027 \times 931.5\,\mathrm{MeV}/c^2 \times c^2 \approx -2.5150\,\mathrm{MeV}.
\end{aligned} \tag{3.20}$$

Finally, the binding energy of the last neutron in $^{16}\mathrm{O}^8$ is

$$\begin{aligned}
\mathrm{B.E.} &= -(M(^{15}\mathrm{O}^8) + m_n - M(^{16}\mathrm{O}^8))c^2 \\
&= -(15.0030 + 1.0087 - 15.9949)\,\mathrm{amu} \times c^2 \\
&\approx -0.0168 \times 931.5\,\mathrm{MeV}/c^2 \times c^2 = -15.6492\,\mathrm{MeV}.
\end{aligned} \tag{3.21}$$

Problem 3.5 *What would you expect for the spin and parity of the ground states of* $^{23}Na^{11}$, $^{35}Cl^{17}$ *and* $^{41}Ca^{20}$ *on the basis of the single-particle shell model? Do these predictions agree with experimental values? What about the magnetic moments of these nuclei? (See CRC Handbook for data.)*

The energy levels in a single-particle shell model are shown in Fig. 3.4 of the text, where the neutrons and protons are assumed to fill the energy levels independently from ground up.

$^{23}\mathrm{Na}^{11}$ has 11 protons and 12 neutrons. The 12 neutrons must all be paired, as are the first 10 protons, and the last proton yields the J^P for the nucleus. The shell structure in this case is

$$\begin{aligned} n &: (1S_{1/2})^2(1P_{3/2})^4(1P_{1/2})^2(1D_{5/2})^4, \\ p &: (1S_{1/2})^2(1P_{3/2})^4(1P_{1/2})^2(1D_{5/2})^3. \end{aligned}$$

Thus, the last proton in the state $1D_{5/2}$ leads to the spin-parity ($\ell = 2$ for the D state)

$$J^P = \frac{5}{2}^+, \tag{3.22}$$

which is not consistent with the observed value of

$$J^P_{\text{obs}} = \frac{3}{2}^+. \tag{3.23}$$

The magnetic moment for this case is given by (see Eq. (3.58) of the text)

$$\mu = 2.79\mu_N + \ell\mu_N = 2.79\mu_N + 2\mu_N = 4.79\mu_N, \tag{3.24}$$

which, not surprisingly, differs from the observed value of

$$\mu_{\text{obs}} = 2.2\mu_N. \tag{3.25}$$

$^{35}\mathrm{Cl}^{17}$ has 17 protons and 18 neutrons. All the neutrons are paired, while 16 of the protons are paired, leaving one unpaired proton, which determines the spin-parity of the nucleus. The shell structure for this case is given by

$$\begin{aligned} n &: (1S_{1/2})^2(1P_{3/2})^4(1P_{1/2})^2(1D_{5/2})^6(2S_{1/2})^2(1D_{3/2})^2, \\ p &: (1S_{1/2})^2(1P_{3/2})^4(1P_{1/2})^2(1D_{5/2})^6(2S_{1/2})^2(1D_{3/2})^1. \end{aligned}$$

The lone proton in the state $^1D_{3/2}$ would lead to

$$J^P = \frac{3}{2}^+, \tag{3.26}$$

which is consistent with experiment. The predicted value of the magnetic moment is

$$\mu = 2.79\mu_N + \ell\mu_N = 2.79\mu_N + 2\mu_N = 4.79\mu_N, \tag{3.27}$$

which does not agree with the observed value of

$$\mu_{\text{obs}} = 0.8\mu_N. \tag{3.28}$$

$^{41}\text{Ca}^{20}$ has 20 paired neutrons, 20 paired protons, and one unpaired neutron. The shell structure has the form

$$n : (1S_{1/2})^2(1P_{3/2})^4(1P_{1/2})^2(1D_{5/2})^6(2S_{1/2})^2(1D_{3/2})^4(1F_{7/2})^1,$$
$$p : (1S_{1/2})^2(1P_{3/2})^4(1P_{1/2})^2(1D_{5/2})^6(2S_{1/2})^2(1D_{3/2})^4.$$

The last neutron in the state $^1F_{7/2}$ determines

$$J^P = \frac{7}{2}^-, \tag{3.29}$$

consistent with experiment. The predicted value of the magnetic moment is that of the unpaired neutron

$$\mu = -1.91\mu_N, \tag{3.30}$$

which differs somewhat from the observed value of

$$\mu_{\text{obs}} = -1.6\mu_N. \tag{3.31}$$

Problem 3.6 *Consider a somewhat more sophisticated model for the anomalous contribution to the magnetic moment of a nucleon. Assume that the proton can be regarded as a fixed neutral center with a π^+ meson circling about in an $\ell = 1$ orbit. Similarly, take a neutron as an effective proton center with a π^- meson in an $\ell = 1$ orbit around it. Using $m_\pi = 140\,\text{MeV}/c^2$, calculate $\mu = \left(\frac{e\hbar}{2m_\pi c}\right)\ell$, and compare results with those of Problem 2.5.*

If we assume such an "atomic" model for the nucleons, then the magnetic moment of the π meson will be given by

$$\mu_\pi = \left(\frac{e\hbar}{2m_\pi c}\right)\ell, \tag{3.32}$$

where e represents the charge of the pion, the mass of the pion is given by

$$m_{\pi^+} = m_{\pi^-} = 140\,\text{MeV}/c^2, \tag{3.33}$$

and ℓ represents the orbital angular momentum of the pion. Since the $\pi^{\pm}$ mesons move in orbits with $\ell = 1$, we obtain

$$\begin{aligned} \mu_{\pi^{\pm}} &= \left(\frac{e\hbar}{2m_{\pi^{\pm}}c}\right) \times 1 = \pm\frac{m_p}{m_{\pi^{\pm}}}\mu_N \\ &= \pm\frac{938.27\,\text{MeV}/c^2}{140\,\text{MeV}/c^2} \times \mu_N \approx \pm 6.7\mu_N. \end{aligned} \tag{3.34}$$

In the model for the proton, where we assume that a π^+ is going around a neutron, we can predict

$$\begin{aligned} \mu_p &= \mu_n + \mu_{\pi^+} \approx (-1.91 + 6.7)\mu_N = 4.79\mu_N \\ &\approx 4.79 \times 3.15 \times 10^{-14}\,\text{MeV/T} \approx 1.51 \times 10^{-13}\,\text{MeV/T}, \end{aligned} \tag{3.35}$$

where we have used the value of μ_N in MeV/T from Eq. (2.30). This is quite comparable to the result in Problem 2.5.

For the neutron, the model assumes that a π^- moves around a stationary proton so that we have

$$\begin{aligned} \mu_n &= \mu_p + \mu_{\pi^-} \approx (2.79 - 6.7)\mu_N = -3.91\mu_N \\ &\approx -3.91 \times 3.15 \times 10^{-14}\,\text{MeV/T} \approx -1.23 \times 10^{-13}\,\text{MeV/T}. \end{aligned} \tag{3.36}$$

Problem 3.7 *The ground state of* $^{137}\text{Ba}^{56}$ *has spin-parity* $\frac{3}{2}^+$. *That is, its spin is* $\frac{3}{2}$ *and parity* $+$. *The first two excited states have spin parity* $\frac{1}{2}^+$ *and* $\frac{11}{2}^-$. *According to the shell model, what assignments would be expected for these excited states? (Hint: The surprise has to do with "pairing energy".)*

$^{137}\text{Ba}^{56}$ has 56 protons and 81 neutrons. The protons are all paired and therefore do not contribute to the spin parity. According to the single-particle shell model, the neutrons should fill the energy levels as follows:

$$\begin{aligned} n: & (1S_{1/2})^2(1P_{3/2})^4(1P_{1/2})^2(1D_{5/2})^6(2S_{1/2})^2(1D_{3/2})^4 \\ & (1F_{7/2})^8(2P_{3/2})^4(1F_{5/2})^6(2P_{1/2})^2(1G_{9/2})^{10}(1G_{7/2})^8 \\ & (2D_{5/2})^6(2D_{3/2})^4(3S_{1/2})^2(1H_{11/2})^{11}. \end{aligned} \tag{3.37}$$

If this is how the energy levels are filled, then the unpaired neutron will be in the state $(1H_{11/2})$, leading to a ground state spin-parity ($\ell = 5$ for an H state)

$$J^P(\text{ground}) = \frac{11}{2}^- . \tag{3.38}$$

Furthermore, the next energy level in the sequence would correspond to the $(1H_{9/2})$ state, so that if the unpaired neutron is excited, then it could move to the next level, leading to a spin-parity assignment for the excited state of

$$J^P(\text{excited}) = \frac{9}{2}^- . \tag{3.39}$$

However, the observed spin-parity of the ground state of $^{137}\text{Ba}^{56}$ is

$$J^P_{\text{obs}} = \frac{3}{2}^+ , \tag{3.40}$$

while the next two excited states are

$$J^P_{\text{obs}}(\text{excited}) = \frac{1}{2}^+ , \frac{11}{2}^- . \tag{3.41}$$

This suggests that the shell structure for the neutrons in $^{137}\text{Ba}^{56}$ may have the form

$$\begin{aligned} n: & (1S_{1/2})^2(1P_{3/2})^4(1P_{1/2})^2(1D_{5/2})^6(2S_{1/2})^2(1D_{3/2})^4 \\ & (1F_{7/2})^8(2P_{3/2})^4(1F_{5/2})^6(2P_{1/2})^2(1G_{9/2})^{10}(1G_{7/2})^8 \\ & (2D_{5/2})^6(2D_{3/2})^3(3S_{1/2})^2(1H_{11/2})^{12}. \end{aligned} \tag{3.42}$$

Since the unpaired neutron is in the state $(2D_{3/2})$, it will lead to a ground state spin-parity assignment of

$$J^P(\text{ground}) = \frac{3}{2}^+ . \tag{3.43}$$

Furthermore, since the filled $(3S_{1/2})$ and $(1H_{11/2})$ states are very close in energy to the $(2D_{3/2})$ state, a neutron from either can "drop down" to fill the $(2D_{3/2})$ shell, which can then yield excited states of spin-parity

$$J^P(\text{excited}) = \frac{1}{2}^+ \quad \text{and} \quad \frac{11}{2}^- . \tag{3.44}$$

We can understand this kind of a shell structure in the following way. While normally energy levels fill from ground up, sometimes it leads to greater nuclear stability when a subshell of higher ℓ is closed off (completely filled) and the unpaired nucleon is left in a lower shell. This happens when shells are very close to each other, as is the case for $3S_{1/2}$, $2D_{3/2}$ and $1H_{11/2}$. For the ground state of ^{137}Ba, the $(2S_{1/2})$ and $(1H_{11/2})$ levels are filled to capacity, while $2D_{3/2}$ is only partially filled, with the unpaired neutron also in that $2D_{3/2}$ state.

4. Nuclear Radiation

Problem 4.1 *Calculate the Q values for the following α-decays between ground-state levels of the nuclei:* (a) $^{208}Po \rightarrow {}^{204}Pb + \alpha$ *and* (b) $^{230}Th \rightarrow {}^{226}Ra + \alpha$. *What are the kinetic energies of the α-particles and of the nuclei in the final state if the decays proceed from rest?*

From the *CRC Handbook* we have the atomic masses

$$\begin{aligned} M(^{208}\mathrm{Po}^{84}) &= 207.9812\,\mathrm{amu}, \quad M(^{204}\mathrm{Pb}^{82}) = 203.9730\,\mathrm{amu}, \\ M(^{230}\mathrm{Th}^{90}) &= 230.0331\,\mathrm{amu}, \quad M(^{226}\mathrm{Ra}^{88}) = 226.0254\,\mathrm{amu}, \\ M(^{4}\mathrm{He}^{2}) &= 4.0026\,\mathrm{amu}. \end{aligned} \tag{4.1}$$

From Eq. (4.4) of the text, the Q value in a reaction involving α decay is given by

$$Q = T_D + T_\alpha = (M(A,Z) - M(A-4,Z-2) - M(4,2))c^2, \tag{4.2}$$

where we assume that $M(A,Z)$ and $M(A-4,Z-2)$ represent the masses of the parent and the daughter nuclei (atomic masses can be used because the masses of the electrons cancel out). Furthermore, the kinetic energies of the α particle and the daughter nuclei are

$$T_\alpha = \frac{M_D}{M_D + M_\alpha} = \frac{M(A-4,Z-2)}{M(A-4,Z-2) + M(4,2)}, \quad T_D = Q - T_\alpha. \tag{4.3}$$

With all this information, we can look at the reaction

$$^{208}\mathrm{Po}^{84} \rightarrow {}^{204}\mathrm{Pb}^{82} + \alpha, \tag{4.4}$$

and we have

$$\begin{aligned}
Q &= \left(M(^{208}\text{Po}^{84}) - M(^{204}\text{Pb}^{82}) - M(^{4}\text{He}^{2})\right) c^2 \\
&\approx (207.9812 - 203.9730 - 4.0026)\,\text{amu} \times c^2 \\
&\approx 0.0056 \times 931.5\,\text{MeV}/c^2 \times c^2 = 5.2164\,\text{MeV}, \\
T_\alpha &= \frac{M(^{204}\text{Pb}^{82})}{M(^{204}\text{Pb}^{82}) + M(^{4}\text{He}^{2})} \\
&\approx \frac{203.9730\,\text{amu}}{(203.9730 + 4.0026)\,\text{amu}} \times 5.2164\,\text{MeV} \\
&\approx 0.98 \times 5.2164\,\text{MeV} \approx 5.11\,\text{MeV}, \\
T_D &= Q - T_\alpha \approx (5.2164 - 5.11)\,\text{MeV} \approx 0.11\,\text{MeV},
\end{aligned} \tag{4.5}$$

where in the intermediate steps we have used Eq. (2.11), which relates the "amu" unit to the "MeV" unit.

Similarly, for the reaction

$$^{230}\text{Th}^{90} \rightarrow {}^{226}\text{Ra}^{88} + \alpha, \tag{4.6}$$

we have

$$\begin{aligned}
Q &= \left(M(^{230}\text{Th}^{90}) - M(^{226}\text{Ra}^{88}) - M(^{4}\text{He}^{2})\right) c^2 \\
&\approx (230.0331 - 226.0254 - 4.0026)\,\text{amu} \times c^2 \\
&\approx 0.0051 \times 931.5\,\text{MeV}/c^2 \times c^2 \approx 4.7506\,\text{MeV}, \\
T_\alpha &= \frac{M(^{226}\text{Ra}^{88})}{M(^{226}\text{Ra}^{88}) + M(^{4}\text{He}^{2})} \\
&\approx \frac{226.0254\,\text{amu}}{(226.0254 + 4.0026)\,\text{amu}} \times 4.7506\,\text{MeV} \\
&\approx 0.982 \times 4.7506\,\text{MeV} \approx 4.66\,\text{MeV}, \\
T_D &= Q - T_\alpha \approx (4.7506 - 4.66)\,\text{MeV} \approx 0.09\,\text{MeV}.
\end{aligned} \tag{4.7}$$

Problem 4.2 *Estimate the relative contribution of the centrifugal barrier and the Coulomb barrier in the scattering of a* 4 MeV *α-particle from ^{236}U. In particular, consider impact parameters of $b = 1$ fm and $b = 7$ fm. What are the orbital quantum numbers in such collisions. (Hint: $|\vec{L}| \sim |\vec{r} \times \vec{p}| \sim \hbar kb \sim \hbar\ell$.)*

The scattering of an α particle from a $^{236}\text{U}^{92}$ nucleus is governed by a Schrödinger equation of the kind given in Eq. (3.28) of the text, with

$$V(r) = V_{\text{Coulomb}} = \frac{ZZ'e^2}{r} = \frac{2 \times 92e^2}{r} = \frac{184e^2}{r},$$
$$V_{\text{centrifugal}} = \frac{\hbar^2 \ell(\ell+1)}{2mr^2}, \tag{4.8}$$

where m is the reduced mass of the system defined in Eq. (1.46) of the text. Since $m_\alpha \ll m_{^{236}\text{U}}$, it follows that

$$m \approx m_\alpha \approx 4.0026\,\text{amu} \approx 4.0026 \times 931.5\,\text{MeV}/c^2 \approx 3728\,\text{MeV}/c^2. \tag{4.9}$$

In the present case, both the Coulomb and the centrifugal terms are repulsive, and while the Coulomb potential is independent of ℓ (acts the same way for every ℓ component), the centrifugal barrier depends explicitly on ℓ. It vanishes for $\ell = 0$ (there is no centrifugal contribution if there is no impact parameter: $\ell \approx |\vec{p} \times \vec{b}|$), and it grows with ℓ. The two potentials have different dependence on radial distance, and must therefore be compared with care. However, qualitatively, for $\ell \neq 0$, the centrifugal potential dominates at very small distances, while the Coulomb potential takes over at large distances. This transition depends on ℓ, and can be seen from Eq. (4.8) to correspond to

$$\begin{aligned} r_{\text{transition}}(\ell) &= \frac{\hbar^2 \ell(\ell+1)}{2mZZ'e^2} = \frac{\hbar c}{2ZZ'mc^2} \times \frac{\hbar c}{e^2} \times \ell(\ell+1) \\ &\approx \frac{197\,\text{MeV} - \text{F}}{2 \times 2 \times 92 \times 3728\,\text{MeV}} \times 137 \times \ell(\ell+1) \\ &\approx 0.0197\ell(\ell+1)\,\text{F}. \end{aligned} \tag{4.10}$$

Classically, any impact parameter b determines the orbital angular momentum in the scattering through Eq. (1.14) of the text

$$\frac{1}{b^2} = \frac{2mE}{\ell^2} \quad \text{or} \quad \ell = \sqrt{2mE}\, b. \tag{4.11}$$

We have

$$m \approx m_\alpha, \quad E = 4\,\text{MeV}. \tag{4.12}$$

Furthermore, since the quantum eigenvalues of angular momentum are represented as $\hbar\ell, \ell = 0, 1, 2, \ldots$, we obtain

$$\begin{aligned}\hbar\ell(b) &\approx \sqrt{2m_\alpha E}\, b = \sqrt{2m_\alpha c^2 E} \times \frac{\hbar b}{\hbar c} \\ &\approx \sqrt{2 \times 3728 \times 4\,(\text{MeV})^2} \times \frac{\hbar b}{197\,\text{MeV} - \text{F}} \\ \text{or}\quad \ell(b) &\approx 0.87 \times \frac{b}{\text{F}}.\end{aligned} \tag{4.13}$$

Therefore, we see that

$$\begin{aligned}\ell(b = 1\,\text{F}) &\approx 0.87 \approx 1, \\ \ell(b = 7\,\text{F}) &\approx 0.87 \times 7 \approx 6.\end{aligned} \tag{4.14}$$

To estimate the relative contributions of the two potentials at the distance of closest approach, we recall that this can be determined from Eq. (1.25) of the text

$$r_0 = \frac{ZZ'e^2}{2E}\left[1 + \sqrt{1 + \left(\frac{2E}{ZZ'e^2}\right)^2 b^2}\,\right]. \tag{4.15}$$

Here, we have

$$\begin{aligned}\frac{ZZ'e^2}{2E} &= \frac{ZZ'\hbar c}{2E} \times \frac{e^2}{\hbar c} \approx \frac{2 \times 92 \times 197\,\text{MeV} - \text{F}}{2 \times 4\,\text{MeV}} \times \frac{1}{137} \\ &\approx 33.12\,\text{F},\end{aligned} \tag{4.16}$$

so that

$$\begin{aligned}r_0(b) &\approx 33.12\ \text{F}\left[1 + \sqrt{1 + \frac{b^2}{(33.12\,\text{F})^2}}\,\right] \\ &\approx 33.12\,\text{F}\left[1 + \left(1 + \frac{b^2}{2(33.12\,\text{F})^2}\right)\right] \\ &\approx 66.24\,\text{F} \approx 66\,\text{F}.\end{aligned} \tag{4.17}$$

(The b-dependent correction terms for the two cases under consideration are negligible, and so, for these b values, r_0 does not depend very much on b.)

At the distance of closest approach, we note from Eq. (4.8) that

$$\begin{aligned}
&V_{\text{Coulomb}}(r = 66\,\text{F}) \\
&\quad\approx \frac{184e^2}{66\,\text{F}} = 184 \times \frac{\hbar c}{66\,\text{F}} \times \frac{e^2}{\hbar c} \\
&\quad\approx 184 \times \frac{197\,\text{MeV} - \text{F}}{66\,\text{F}} \times \frac{1}{137} \approx 4\,\text{MeV}, \\
&V_{\text{centrifugal}}(r = 66\,\text{F}, \ell = 6) \qquad\qquad (4.18) \\
&\quad\approx \frac{\hbar^2\ell(\ell+1)}{2m_\alpha r^2} = \frac{\hbar c}{2m_\alpha c^2} \times \frac{\hbar c}{r^2} \times \ell(\ell+1) \\
&\quad\approx \frac{197\,\text{MeV} - \text{F}}{2 \times 3728\,\text{MeV}} \times \frac{197\,\text{MeV} - \text{F}}{(66\,\text{F})^2} \times 6 \times 7 \\
&\quad\approx 0.05\,\text{MeV}.
\end{aligned}$$

We chose $\ell = 6$ for the calculation of the centrifugal term because it is larger for higher ℓ values, and yet we find that it is negligible compared to the Coulomb potential for $\ell = 6$, even at the distance of closest approach. This is easily understood from our earlier observation that

$$r_{\text{transition}}(\ell = 6) \approx 0.0197 \times 6 \times 7\,\text{F} \approx 0.83\,\text{F} \ll r_0, \qquad (4.19)$$

as a result of which, the Coulomb potential always dominates over the centrifugal term for these cases.

Problem 4.3 *Free neutrons decay into protons, electrons and antineutrinos, with a mean life of* 889 sec. *If the neutron-proton mass difference is taken as* $1.3\,\text{MeV}/c^2$, *calculate to at least* 10% *accuracy the maximum kinetic energies that electrons and protons can have. What would be the maximum energy that the antineutrinos can have? (Assume decay from rest and that the antineutrino is massless.)*

The decay of a free neutron leads to three bodies in the final state:

$$n \to p + e^- + \bar{\nu}_e, \qquad (4.20)$$

and, as discussed in the text, unlike α decay, which constitutes a two-body decay, the energies of the individual decay products cannot be specified uniquely. If we assume that the neutron decays at rest, then momentum conservation leads to

$$\vec{p}_p + \vec{p}_e + \vec{p}_{\bar{\nu}} = 0, \tag{4.21}$$

and it follows from the conservation of energy in Eq. (4.32) of the text that

$$Q = T_p + T_e + T_{\bar{\nu}} = (m_n - m_p - m_e - m_{\bar{\nu}})c^2. \tag{4.22}$$

We are given that

$$(m_n - m_p)c^2 = 1.3\,\text{MeV}, \quad m_e c^2 = 0.511\,\text{MeV}, \quad m_{\bar{\nu}} = 0, \tag{4.23}$$

it therefore follows from (4.22) that

$$Q = T_p + T_e + T_{\bar{\nu}} = (1.3 - 0.511 - 0)\,\text{MeV} \approx 0.8\,\text{MeV}. \tag{4.24}$$

This relationship shows why the energies of the individual decay products cannot in general be unique. However, if one of the decay products is at rest (namely, one of the kinetic energies vanishes), then conservation of momentum provides a way to determine the unique energies of the other two products. Under these circumstances, one of the objects can assume its maximum energy. There are three such special cases to consider, and we will analyze these separately in what follows.

First, if the antineutrino is produced at rest (since the antineutrino is assumed to be massless, it cannot really be produced at rest, and this should therefore be interpreted as saying that the momentum and, consequently, the energy of the antineutrino is negligible), it follows from Eqs. (4.21) and (4.22) that

$$\vec{p}_p = -\vec{p}_e, \quad T_p + T_e = Q. \tag{4.25}$$

Since the sum of the kinetic energies is $\approx 0.8\,\text{MeV}$, the proton must be nonrelativistic, whereas the electron can be relativistic. Therefore, using the appropriate forms for kinetic energies, we obtain

$$\begin{aligned} \frac{p_p^2}{2m_p} + \sqrt{p_e^2c^2 + m_e^2c^4} - m_ec^2 = Q \\ \text{or} \quad \sqrt{p_e^2c^2 + m_e^2c^4} = Q + m_ec^2 - \frac{p_e^2}{2m_p}, \end{aligned} \tag{4.26}$$

where we have used the identification in Eq. (4.25). Squaring both sides, we can solve for p_e^2

$$\begin{aligned} p_e^2c^2 + m_e^2c^4 &= \left(Q + m_ec^2\right)^2 + \frac{(p_e^2)^2}{4m_p^2} - \frac{p_e^2}{m_p(Q + m_ec^2)} \\ \text{or} \quad (p_e^2)^2 &- 4m_pp_e^2\left(Q + (m_p + m_e)c^2\right) + 4m_p^2\left(Q + 2m_ec^2\right) = 0 \\ \text{or} \quad p_e^2 &= 2m_p\left[Q + (m_p + m_e)c^2 \pm \sqrt{(m_p + m_e)^2c^4 + 2Qm_pc^2}\right]. \end{aligned} \tag{4.27}$$

Since $m_pc^2 \gg m_ec^2$, and $m_pc^2 \gg Q$, we can approximate the kinetic energy of the proton as

$$\begin{aligned} T_p = \frac{p_e^2}{2m_p} &\approx Q + (m_p + m_e)c^2 \pm \left((m_p + m_e)c^2 + \frac{Qm_pc^2}{(m_p + m_e)c^2}\right) \\ &= Q\left(1 - \frac{m_pc^2}{(m_p + m_e)c^2}\right) = \frac{m_ec^2}{(m_p + m_e)c^2}Q \\ &= \frac{0.511\,\text{MeV}}{(938.27 + 0.511)\,\text{MeV}} \times 0.8\,\text{MeV} \approx 4.24 \times 10^{-4}\,\text{MeV} \\ &\approx 0.4\,\text{keV}, \end{aligned} \tag{4.28}$$

where in the intermediate steps we have discarded the solution of the quadratic equation with positive sign as unphysical, since it leads to a larger value of proton kinetic energy than is allowed by the bound in Eq. (4.24). The kinetic energy of the electron is therefore

$$T_e = Q - T_p \approx 0.8\,\text{MeV} - 0.4\,\text{keV} = 0.7996\,\text{MeV}. \tag{4.29}$$

The second case to consider is when the electron is emitted at rest, namely, $\vec{p}_e = 0$. Here we have

$$\vec{p}_p = -\vec{p}_{\bar{\nu}}, \quad T_p + T_{\bar{\nu}} + Q. \tag{4.30}$$

For a massless neutrino, $T_{\bar{\nu}} = E_{\bar{\nu}} = p_{\bar{\nu}}c$ (where $p_{\bar{\nu}} = |\vec{p}_{\bar{\nu}}|$), so that we obtain

$$\begin{aligned}
&T_p + T_{\bar{\nu}} = Q \\
\text{or}\quad &\frac{p_{\bar{\nu}}^2}{2m_p} + p_{\bar{\nu}}c = Q \qquad (4.31)\\
\text{or}\quad &p_{\bar{\nu}} = -m_p c \pm \sqrt{m_p^2 c^2 + 2m_p Q} \approx \frac{Q}{c},
\end{aligned}$$

where we have discarded the solution with the negative sign because it leads to a negative value for $p_{\bar{\nu}}$, which is unphysical. The kinetic energies of the proton and the neutrino now become

$$\begin{aligned}
T_p &= \frac{p_{\bar{\nu}}^2}{2m_p} \approx \frac{Q^2}{2m_p c^2} \\
&\approx \frac{(0.8\,\text{MeV})^2}{2 \times 938\,\text{MeV}} \approx 3.39 \times 10^{-4}\,\text{MeV} = 0.34\,\text{keV}, \qquad (4.32)\\
T_{\bar{\nu}} &= Q - T_p \approx 0.8\,\text{MeV} - 0.34\,\text{keV} = 0.7996\,\text{MeV}.
\end{aligned}$$

The final special case is when the proton is produced at rest, namely, when $\vec{p}_p = 0$. Here, we have

$$\vec{p}_e = -\vec{p}_{\bar{\nu}}, \quad T_e + T_{\bar{\nu}} = Q. \qquad (4.33)$$

Using these relationships we obtain

$$\begin{aligned}
&T_e + T_{\bar{\nu}} = \sqrt{p_{\bar{\nu}}^2 c^2 + m_e^2 c^4} - m_e c^2 + p_{\bar{\nu}}c = Q \\
\text{or}\quad &\sqrt{p_{\bar{\nu}}^2 c^2 + m_e^2 c^4} = Q + m_e c^2 - p_{\bar{\nu}}c.
\end{aligned} \qquad (4.34)$$

Squaring both sides and rearranging terms, we obtain

$$\begin{aligned}
p_{\bar{\nu}} &= \frac{Q^2 + 2Qm_e c^2}{2(Q + m_e c^2)c}, \\
T_{\bar{\nu}} &= p_{\bar{\nu}}c = \frac{Q^2 + 2Qm_e c^2}{2(Q + m_e c^2)} = \frac{0.8\,\text{MeV}(0.8 + 2 \times 0.511)\,\text{MeV}}{2 \times (0.8 + 0.511)\,\text{MeV}} \qquad (4.35)\\
&\approx 0.5\,\text{MeV}, \\
T_e &= Q - T_{\bar{\nu}} \approx 0.8\,\text{MeV} - 0.5\,\text{MeV} = 0.3\,\text{MeV}.
\end{aligned}$$

Comparing Eqs. (4.28), (4.29), (4.33) and (4.35), we conclude that

$$\begin{aligned} T_p^{(\max)} &\approx 0.4\,\text{keV}, && \text{when the neutrino is at rest,} \\ T_e^{(\max)} &\approx 0.7996\,\text{MeV}, && \text{when the neutrino is at rest,} \\ T_{\bar{\nu}}^{(\max)} &\approx 0.7996\,\text{MeV}, && \text{when the electron is at rest.} \end{aligned} \tag{4.36}$$

Problem 4.4 *If the stable isotope of sodium is ^{23}Na, what kind of radioactivity would you expect from* (a) ^{22}*Na and* (b) ^{24}*Na?*

We know that $^{23}\text{Na}^{11}$ is stable. The isotope $^{22}\text{Na}^{11}$ has one less neutron, while $^{24}\text{Na}^{11}$ has one extra neutron relative to $^{23}\text{Na}^{11}$. Consequently, a proton in $^{22}\text{Na}^{11}$ can undergo an inverse β decay to yield

$$^{22}\text{Na}^{11} \to {}^{22}\text{Ne}^{10} + e^+ + \nu_e, \tag{4.37}$$

where $^{22}\text{Ne}^{10}$ is a naturally occurring stable isotope of $^{20}\text{Ne}^{10}$. Similarly, the extra neutron in $^{24}\text{Na}^{11}$ can undergo a β decay to yield

$$^{24}\text{Na}^{11} \to {}^{24}\text{Mg}^{12} + e^- + \bar{\nu}_e, \tag{4.38}$$

where $^{24}\text{Mg}^{12}$ is stable.

Problem 4.5 *Specify any additional particles needed in the following weak reactions to assure the conservation of lepton number:* (a) $\mu^- \to e^- +\ ?$ (b) $\tau^+ \to e^+\ ?$ (c) $e^- + {}^AX^Z \to\ ?$ (d) $\nu_\mu + n \to\ ?$ (e) $^AX^Z \to {}^AY^{Z-1} +\ ?$ (f) $\bar{\nu}_e + p \to\ ?$

As discussed in Chap. 4 of the text, all leptons carry a quantum number known as *lepton number.* If the electron, the muon and the tau lepton carry the same lepton number 1, then a reaction such as

$$\mu^- \to e^- + \gamma, \tag{4.39}$$

would be kinematically allowed, and would satisfy lepton-number conservation. However, as discussed in Chap. 9 (see Sec. 9.2.2 of the text), different families of leptons carry different lepton-family quantum numbers. If these quantum numbers are to be conserved in a reaction, then we can have only the following unique reactions

$$
\begin{aligned}
&\text{(a)} \quad \mu^- \to e^- + \bar{\nu}_e + \nu_\mu, \\
&\text{(b)} \quad \tau^+ \to e^+ + \nu_e + \nu_\tau, \\
&\text{(c)} \quad e^- + {}^A\mathrm{X}^Z \to {}^A\mathrm{Y}^{Z-1} + \nu_e, \\
&\text{(d)} \quad \nu_\mu + n \to p + \mu^-, \\
&\text{(e)} \quad {}^A\mathrm{X}^Z \to {}^A\mathrm{Y}^{Z-1} + e^+ + \nu_e, \\
&\text{(f)} \quad \bar{\nu}_e + p \to e^+ + n.
\end{aligned} \tag{4.40}
$$

Problem 4.6 *Calculate the typical kinetic energy expected of an α-particle confined within a nucleus if its emitted energy is* 10 MeV. *What is the momentum of such an α-particle inside the nucleus and after it is emitted. Is the wavelength of such an α-particle acceptable for it to be contained within a nucleus of ^{12}C? What about ^{238}U?*

An α particle with 10 MeV of kinetic energy is clearly nonrelativistic. Outside the nuclear potential well, all its energy is kinetic and we can therefore identify

$$
T_\alpha^{(\text{outside})} = E = 10\,\text{MeV}. \tag{4.41}
$$

On the other hand, inside the nuclear well, the α particle feels the potential of the well and we have

$$
E = T_\alpha^{(\text{inside})} + V(r) = T_\alpha^{(\text{inside})} - U_0, \tag{4.42}
$$

where the potential depth of the nuclear well (if we assume a square-well potential) is about $U_0 = 40\,\text{MeV}$ (as discussed in Sec. 4.3 of the text). Therefore, using Eq. (4.41) and conservation of energy, we obtain

$$
\begin{aligned}
E = T_\alpha^{(\text{inside})} - U_0 &= 10\,\text{MeV} \\
T_\alpha^{(\text{inside})} &= 10\,\text{MeV} + U_0 = (10 + 40)\,\text{MeV} = 50\,\text{MeV}.
\end{aligned} \tag{4.43}
$$

In other words, the α particle has more kinetic energy inside the nucleus than outside.

Since the α particle is nonrelativistic both inside and outside the potential well, we can calculate its momentum as

$$\begin{aligned}
p^{(\text{outside})} &= \sqrt{2m_\alpha T_\alpha^{(\text{outside})}} = \sqrt{2m_\alpha c^2 E} \times \frac{1}{c} \\
&\approx \sqrt{2 \times 3728\,\text{MeV} \times 10\,\text{MeV}} \times \frac{1}{c} \\
&\approx 273\,\text{MeV}/c, \\
p^{(\text{inside})} &= \sqrt{2m_\alpha T_\alpha^{(\text{inside})}} = \sqrt{2m_\alpha c^2 T_\alpha^{(\text{inside})}} \times \frac{1}{c} \\
&\approx \sqrt{2 \times 3728\,\text{MeV} \times 50\,\text{MeV}} \times \frac{1}{c} \\
&\approx 610\,\text{MeV}/c,
\end{aligned} \tag{4.44}$$

where we have used the value of m_α given in Eq. (4.9).

The corresponding de Broglie wavelength of the α particle inside the nuclear well is

$$\begin{aligned}
\lambda^{(\text{inside})} &= \frac{h}{p^{(\text{inside})}} = \frac{2\pi\hbar c}{p^{(\text{inside})}c} \\
&\approx \frac{6 \times 197\,\text{MeV} - \text{F}}{610\,\text{MeV}} \approx 1.9\,\text{F} \\
&= 1.9 \times 10^{-13}\,\text{cm},
\end{aligned} \tag{4.45}$$

where we have used the value of the momentum from Eq. (4.44).

From Eq. (2.16) of the text, we can obtain

$$\begin{aligned}
R_{^{12}\text{C}} &= 1.2 \times 10^{-13}\,A^{\frac{1}{3}}\,\text{cm} = 1.2 \times (12)^{\frac{1}{3}} \times 10^{-13}\,\text{cm} \\
&\approx 2.76 \times 10^{-13}\,\text{cm} > \lambda^{(\text{inside})}, \\
R_{^{238}\text{U}} &= 1.2 \times (238)^{\frac{1}{3}} \times 10^{-13}\,\text{cm} \\
&\approx 7.44 \times 10^{-13}\,\text{cm} > \lambda^{(\text{inside})}.
\end{aligned} \tag{4.46}$$

We therefore conclude that the α particle can be contained inside either of these nuclei.

Problem 4.7 *When you examine the dependence of Z on N for stable nuclei, you find that β^+ emitters lie above the region of stability (have proton excess) and β^- emitters lie below that region (have* *neutron excess). For example, ^8B emits β^+, while ^{12}B emits β^-. Stable nuclei are those that do not seem to have sufficient mass for either emission to take place, that is, they are the nuclei with greatest binding or smallest mass. As discussed in Problem 3.1, this suggests that stable nuclei should correspond to a "valley" in the M-Z space, that is, specified by $\frac{\partial M}{\partial Z} = 0$. Using the semi-empirical mass formula for M, show that the relationship between Z and A for this valley of stability is $Z \approx \frac{A}{(2+0.015\,A^{2/3})}$. Several nuclei with Z beyond 110 were discovered in the late 1990s. Is it possible that there could be more "islands" of stability for $Z > 120$? Consider, for specifics, the possibility of binding of $Z = 125$, $Z = 126$, and $Z = 164$. Even more massive nuclei have been hypothesized with $Z > 200$. These would have rather exotic bubble-like or toroidal structure. Why would such structures be expected to be more stable than spherical nuclei?*

As derived in Problem 3.1 (see Eq. (3.2)), the valley for stable nuclei can be described as

$$Z = \frac{A}{2}\,\frac{4a_4 + (m_n - m_p)c^2}{4a_4 + a_3 A^{\frac{2}{3}}}. \tag{4.47}$$

Using the values (see Eqs. (2.2) and (3.4) of the text)

$$(m_n - m_p)c^2 \approx 1.3\,\text{MeV}, \quad a_3 \approx 0.72\,\text{MeV}, \quad a_4 \approx 23.3\,\text{MeV}, \tag{4.48}$$

we obtain

$$\begin{aligned} Z &\approx \frac{A}{2} \times \frac{(4 \times 23.3 + 1.3)\,\text{MeV}}{(4 \times 23.3 + 0.72 A^{\frac{2}{3}})\,\text{MeV}} = \frac{A}{2} \times \frac{(93.2 + 1.3)}{(93.2 + 0.72 A^{\frac{2}{3}})} \\ &\approx \frac{A}{2} \times \frac{93.2}{93.2 + 0.72 A^{\frac{2}{3}}} \approx \frac{A}{2 + 0.015 A^{\frac{2}{3}}}. \end{aligned} \tag{4.49}$$

(Note that the coefficient of the second term in the denominator may have a typo in older edition of the text.)

$Z = 125$ is an unfavored odd number, and not likely to be stable. Shell Model corrections to the semi-empirical mass formula would favor stability for $Z = 126$ and $A = 310$. However, here $Z^2/A = 51$, which is above the $Z^2/A = 47$ limit, where spontaneous fission should prevent formation of stable nuclei. A toroidal shape would permit the protons to separate (as in fission), thereby lowering the impact of the destabilizing Coulomb term in the mass formula.

5. Applications of Nuclear Physics

Problem 5.1 *To study neutron absorption cross sections at very low energies, one must often slow down (moderate) energetic* ($\approx 1\,\mathrm{MeV}$) *neutrons that are produced in reactors. Show that paraffin would be a better moderator than aluminum, by specifically calculating the maximum energy that a* $1\,\mathrm{MeV}$ *neutron can transfer in a collision with a proton (within paraffin) as opposed to that with an Al nucleus.*

As we saw in Problem 2.9 (see Eq. (2.48)), in a head-on collision with a target nucleus of mass number A, a neutron scatters backwards ($\theta = \pi$) with an energy

$$E_n = \left(\frac{A-1}{A+1}\right)^2 E_0, \tag{5.1}$$

where E_0 represents the energy of the incident neutron. If the target is paraffin (essentially a chain of CH_2 units), then ignoring the presence of C, for $A = 1$ we get

$$E_n^{(\text{paraffin})} = \left(\frac{A-1}{A+1}\right)^2 E_0 = \left(\frac{1-1}{1+1}\right)^2 E_0 = 0. \tag{5.2}$$

On the other hand, if the target nucleus is aluminum ($^{27}\mathrm{Al}^{13}$), we have $A = 27$, leading to

$$\begin{aligned} E_n^{(\text{aluminum})} &= \left(\frac{A-1}{A+1}\right)^2 E_0 = \left(\frac{27-1}{27+1}\right)^2 E_0 \\ &\approx 0.86 E_0. \end{aligned} \tag{5.3}$$

Thus, we see that paraffin is a much better moderator, since after one collision a neutron can lose all its energy.

Problem 5.2 *Calculate the energy released when* 1 gm *of* ^{235}U *fissions into* ^{148}La *and* ^{87}Br. *Compare this to the energy released in fusing deuterium and tritium nuclei in* 1 gm *of tritiated water with* 1 gm *of deuterated water (i.e.* T_2O *and* D_2O*).*

Let us consider the fission process

$$^{235}\mathrm{U}^{92} + n \to {}^{148}\mathrm{La}^{57} + {}^{87}\mathrm{Br}^{35} + n. \tag{5.4}$$

The energy released per fission is

$$\Delta = \mathrm{B.E.}(^{235}\mathrm{U}^{92}) - \mathrm{B.E.}(^{148}\mathrm{La}^{57}) - \mathrm{B.E.}(^{87}\mathrm{Br}^{35}). \tag{5.5}$$

The B.E. for the different nuclei can be calculated using the phenomenological formula Eq. (3.3) of the text, or more directly from the defining relationship in Eq. (2.5) of the text. Using the latter, we obtain

$$\begin{aligned}\Delta &= [M(^{235}\mathrm{U}^{92}) - 92m_p - 143m_n - M(^{148}\mathrm{La}^{57}) + 57m_p + 91m_n \\ &\quad - M(^{87}\mathrm{Br}^{35}) + 35m_p + 52m_n]c^2 \\ &= [M(^{235}\mathrm{U}^{92}) - M(^{148}\mathrm{La}^{57}) - M(^{87}\mathrm{Br}^{35})]c^2. \end{aligned} \tag{5.6}$$

From the *CRC Handbook*, we have the atomic masses

$$\begin{aligned} M(^{235}\mathrm{U}^{92}) &\approx 235.0439\,\mathrm{amu}, \\ M(^{148}\mathrm{La}^{57}) &\approx 147.9320\,\mathrm{amu}, \\ M(^{87}\mathrm{Br}^{35}) &\approx 86.9207\,\mathrm{amu}. \end{aligned} \tag{5.7}$$

Using these and the conversion factor from "amu" units to "MeV" units in Eq. (2.11), we obtain

$$\begin{aligned}\Delta &\approx (235.0439 - 147.9320 - 86.9207)\ \mathrm{amu} \times c^2 \\ &= 0.1912 \times 931.5\,\mathrm{MeV}/c^2 \times c^2 \approx 178.1\,\mathrm{MeV}. \end{aligned} \tag{5.8}$$

This is the energy released per nuclear fission. In 1 g of ^{235}U, we have

$$N_{^{235}\mathrm{U}} = \frac{6 \times 10^{23}}{235} \tag{5.9}$$

so that the complete fission of 1 g of the material will lead to a release of

$$\begin{aligned} E_{\mathrm{fission}} &= N_{^{235}\mathrm{U}} \times \Delta \approx \frac{6 \times 10^{23}}{235} \times 178.1\,\mathrm{MeV} \\ &\approx 4.55 \times 10^{23}\,\mathrm{MeV}. \end{aligned} \tag{5.10}$$

From Eq. (5.21) of the text, we note that the fusion of a deuterium ($^2H^1$) nucleus with tritium ($^3H^1$) leads to

$$^2H + {}^3H \to {}^4He + n + 17.6\,\text{MeV}. \tag{5.11}$$

Now,

$$A_{D_2O} = 2 \times 2 + 16 = 20, \quad A_{T_2O} = 2 \times 3 + 16 = 22, \tag{5.12}$$

and in 1 g of D_2O, there are

$$N_{D_2O} = 2 \times \frac{6 \times 10^{23}}{20} = 6 \times 10^{22} \tag{5.13}$$

deuterium nuclei, while in 1 g of T_2O, there are

$$N_{T_2O} = 2 \times \frac{6 \times 10^{23}}{22} \approx 5.4 \times 10^{22} \tag{5.14}$$

tritium nuclei. When we combine 1 g of D_2O with 1 g of T_2O, we therefore have

$$N = N_{T_2O} \approx 5.4 \times 10^{22} \tag{5.15}$$

possible number of fusions. Complete fusion will lead to a release of energy amounting to

$$\begin{aligned} E_{\text{fusion}} &= N \times 17.6\,\text{MeV} \approx 5.4 \times 10^{22} \times 17.6\,\text{MeV} \\ &\approx 9.5 \times 10^{23}\,\text{MeV}. \end{aligned} \tag{5.16}$$

Comparing Eqs. (5.10) and (5.16), we conclude that there will be a comparable amount of energy released in fission and fusion processes per gram of material.

Problem 5.3 *The counting rate for a radioactive source is measured for one minute intervals every hour, and the resulting counts are:* $107, 84, 65, 50, 36, 48, 33, 25, \ldots$. *Plot the counting rate versus time, and from the graph roughly estimate the mean life and the half-life. Recalling that the expected error on* N *counts is* $\sqrt{N}$, *do the data points seem reasonable? (Hint: use "semi-log" paper to plot* $\log N$ *vs.* t.*)*

The activity of a radioactive material is given in Eq. (5.26) of the text

$$\mathcal{A}(t) = \lambda N_0 e^{-\lambda t} = \mathcal{A}(0)e^{-\lambda t}, \tag{5.17}$$

where the decay constant λ is related to the half-life of the material. The activity gives the number of nuclear disintegrations per second at some time t. We can find the number of decays within a time interval Δt centered on t, as follows:

$$\begin{aligned}\Delta N(t) &= \int_{t-\frac{\Delta t}{2}}^{t+\frac{\Delta t}{2}} dt'\, \mathcal{A}(t') = \mathcal{A}(0) \int_{t-\frac{\Delta t}{2}}^{t+\frac{\Delta t}{2}} dt'\, e^{-\lambda t'} \\ &= -\frac{1}{\lambda}\, \mathcal{A}(0) \left(e^{-\lambda(t+\frac{\Delta t}{2})} - e^{-\lambda(t-\frac{\Delta t}{2})}\right) \\ &= \tau \mathcal{A}(0)e^{-\lambda t} \left(e^{\frac{\lambda \Delta t}{2}} - e^{-\frac{\lambda \Delta t}{2}}\right),\end{aligned} \tag{5.18}$$

where, as defined in Eqs. (5.24) and (5.25) of the text, the mean life τ is given by

$$\tau = \frac{1}{\lambda} = \frac{t_{1/2}}{\ln 2}. \tag{5.19}$$

When the counting interval is

$$\Delta t = 1\,\text{min} = 60\,\text{sec}, \tag{5.20}$$

we can write

$$\begin{aligned}\Delta N(t) &= \tau \mathcal{A}(0)e^{-\lambda t}\left(e^{30\lambda} - e^{-30\lambda}\right) \\ \text{or} \quad \ln \Delta N(t) &= -\lambda t + C,\end{aligned} \tag{5.21}$$

where we have combined the constant terms into C

$$C = \ln(\tau \mathcal{A}(0)) + \ln(e^{30\lambda} - e^{-30\lambda}). \tag{5.22}$$

Equation (5.21) represents a straight line with slope $(-\lambda)$ and an intercept C. Let us tabulate the observed counting rates (number of counts in 1-minute intervals) at the start of each hour as a function of time, as shown in the table on the next page.

t (hr)	$\Delta N(t)$	$\ln \Delta N(t)$
1	107	≈ 4.67
2	84	≈ 4.43
3	65	≈ 4.17
4	50	≈ 3.91
5	36	≈ 3.58
6	48	≈ 3.87
7	33	≈ 3.50
8	25	≈ 3.22

We can now plot $\ln \Delta N(t)$ against t, and as we have already mentioned, the plot should yield a straight line with slope λ.

In Fig. 5.1, we plot the counting rate as a semilog graph, in which case there is no need to calculate $\ln \Delta N(t)$, and the slope can be

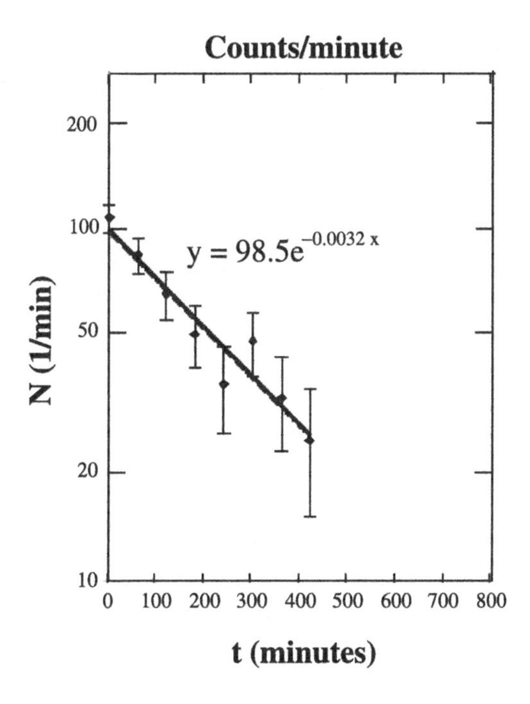

Fig. 5.1. Counts per minute vs. time on semilog paper.

obtained more simply. Note that a change by a factor of 10 in $N(t)$ corresponds to a change of 1 in $\ln \Delta N(t)$. The slope in log (not ln) can therefore be read off the graph! (A least-square fit to the data is also shown in the plot.)

The slope calculated from the data gives an estimate of

$$\lambda \approx 3.1 \times 10^{-3}\,\text{min}^{-1} \tag{5.23}$$

which, in turn, leads to

$$\begin{aligned} \tau &= \text{mean life} = \frac{1}{\lambda} \approx 322\,\text{min}, \\ t_{1/2} &= \text{half life} = \tau \ln 2 \approx 224\,\text{min}. \end{aligned} \tag{5.24}$$

The statistical uncertainties shown in Fig. 5.1 correspond to square roots in the number of events (Poisson statistics). The fit is reliable since 7 of the 8 points lie within one standard deviation (error bar) of the straight line fitted to the data.

Problem 5.4 *A relic from an Egyptian tomb contains* 1 gm *of carbon with a measured activity of* 4×10^{-12} Ci. *If the ratio of* $\frac{^{14}C}{^{12}C}$ *nuclei in a live tree is* 1.3×10^{-12}, *how old is the relic? Assume the half-life of* ^{14}C *is* 5730 yr.

We know from the previous problem (as well as from Eq. (5.26) of the text) that

$$\mathcal{A}(t) = \mathcal{A}(0)e^{-\lambda t}, \quad \mathcal{A}(0) = \lambda N(0), \tag{5.25}$$

where the decay constant λ is related to the half-life as

$$\lambda = \frac{\ln 2}{t_{1/2}} = \frac{0.693}{t_{1/2}}. \tag{5.26}$$

For the present problem, we are given that

$$\begin{aligned} t_{1/2}^{(^{14}\text{C})} &= 5730\,\text{yr} = 5730 \times 365 \times 24 \times 60 \times 60\,\text{sec} \\ &\approx 1.8 \times 10^{11}\,\text{sec}, \\ \lambda^{(^{14}\text{C})} &= \frac{0.693}{t_{1/2}^{^{14}\text{C}}} \approx \frac{0.693}{1.8 \times 10^{11}\,\text{sec}} \\ &\approx 3.8 \times 10^{-12}/\text{sec}. \end{aligned} \tag{5.27}$$

Since the ratio of $\frac{^{14}\mathrm{C}}{^{12}\mathrm{C}}$ nuclei in a living tree is given as 1.3×10^{-12}, in 1 g of carbon, the number of ^{14}C nuclei is given by

$$N_{(^{14}\mathrm{C})} \approx 1.3 \times 10^{-12} \times \frac{6 \times 10^{23}}{12} = 6.5 \times 10^{10}. \qquad (5.28)$$

It follows therefore that

$$\begin{aligned} \mathcal{A}(0) &= \lambda^{(^{14}\mathrm{C})} N_{(^{14}\mathrm{C})}(0) \approx 3.8 \times 10^{-12}/\mathrm{sec} \times 6.5 \times 10^{10}\,\mathrm{decays} \\ &\approx 0.25\,\mathrm{decays/sec}. \end{aligned} \qquad (5.29)$$

The present activity of the relic is

$$\begin{aligned} \mathcal{A}(t) &= 4 \times 10^{-12}\,\mathrm{Ci} = 4 \times 10^{-12} \times 3.7 \times 10^{10}\,\mathrm{decays/sec} \\ &\approx 0.15\,\mathrm{decays/sec}, \end{aligned} \qquad (5.30)$$

where we have used the definition of Curie given in Eq. (5.29) of the text.

Using these values, we determine from Eq. (5.25) that

$$\begin{aligned} -\lambda^{(^{14}\mathrm{C})} t &= \ln \frac{\mathcal{A}(t)}{\mathcal{A}(0)} \approx \ln \frac{0.15}{0.25} \approx -0.51 \\ \text{or} \quad t &\approx \frac{0.51}{\lambda^{(^{14}\mathrm{C})}} \approx \frac{0.51}{3.8 \times 10^{-12}/\mathrm{sec}} \approx 1.3 \times 10^{11}\,\mathrm{sec} \qquad (5.31) \\ &\approx \frac{1.3 \times 10^{11}\,\mathrm{sec}}{3.1 \times 10^{7}\,\mathrm{sec/yr}} \approx 4193\,\mathrm{yrs}. \end{aligned}$$

Thus, the relic is approximately 4193 yrs old.

Problem 5.5 *If the lifetime of the proton is* 10^{33} yr, *how many proton decays would you expect per year in a mass of* 10^3 metric tons *of water? What would be the approximate number expected in the year* 2050?

If the mean life of the proton is

$$\tau_p = 10^{33}\,\mathrm{yr} \approx 10^{33} \times 3.1 \times 10^{7}\,\mathrm{sec} = 3.1 \times 10^{40}\,\mathrm{sec}, \qquad (5.32)$$

then its decay constant is

$$\lambda_p = \frac{1}{\tau_p} = 10^{-33}/\mathrm{yr} \approx \frac{10^{-33}}{3.1 \times 10^{7}\,\mathrm{sec}} \approx 3.2 \times 10^{-41}/\mathrm{sec}. \qquad (5.33)$$

This is extremely small.

The activity, as we have seen in the past two problems, is defined as

$$\mathcal{A}(t) = \mathcal{A}(0)e^{-\lambda_p t}, \tag{5.34}$$

where

$$\mathcal{A}(0) = \lambda_p N_p(0) = 10^{-33} N_p(0)\,\text{decays/yr}. \tag{5.35}$$

The weight of 10^3 metric tons of water (H_2O) is given by

$$10^3 \times 10^3\,\text{kg} = 10^6 \times 10^3\,\text{g} = 10^9\,\text{g}, \tag{5.36}$$

and contains (for $A(H_2O) = 2 \times 1 + 16 = 18$)

$$N_p = 2 \times 10^9 \times \frac{6 \times 10^{23}}{18} \approx 6.7 \times 10^{31} \tag{5.37}$$

protons (nuclei of hydrogen). Therefore, we see that the present activity (assumed as that at $t = 0$) is

$$\begin{aligned}\mathcal{A}(0) &= 10^{-33} N_p(0)\,\text{decays/yr} = 10^{-33} \times 6.7 \times 10^{31}\,\text{decays/yr}\\ &\approx 0.067\,\text{decays/yr}.\end{aligned} \tag{5.38}$$

The number of proton decays in 10^3 metric tons of water per year is negligibly small. Furthermore, from Eq. (5.34)

$$\mathcal{A}(t) = \mathcal{A}(0)e^{-\lambda_p t} = 0.067 \times e^{-10^{-33}t}/\text{yr}, \tag{5.39}$$

we conclude that in the year 2050 ($t \approx 50\,\text{yr}$), the counting rate will not change significantly.

Although we have ignored the decays of bound protons (and neutrons), there is really no reason for doing that. Hence, in principle, the nucleons within oxygen nuclei can also contribute to the decay rate through barrier penetration. Of course, because of large nuclear binding energies, such decays could not proceed via standard β-decay (see Chap. 15 in the text).

Problem 5.6 *Calculate the surface energies and Coulomb energies for the following nuclei:*

$$^{228}Th,\ ^{234}U,\ ^{236}U,\ ^{240}Pu,\ ^{243}Pu.$$

Based on your calculations which nuclei would you expect to fission most easily?

For nuclei that can be described as spherical liquid drops, the contributions of the surface and the Coulomb energies to the B.E. are given in Eq. (3.2) of the text

$$
\begin{aligned}
S(A,Z) &= \text{surface energy} = a_2 A^{\frac{2}{3}} \approx 16.8 A^{\frac{2}{3}}, \\
C(A,Z) &= \text{Coulomb energy} = a_3 \frac{Z^2}{A^{\frac{1}{3}}} \approx 0.72 A^{\frac{2}{3}} \times \frac{Z^2}{A},
\end{aligned}
\tag{5.40}
$$

where we have used the values of a_2 and a_3 given in Eq. (3.4) of the text. We note that both these energies are positive and therefore have a destabilizing effect on any nucleus. Carrying out the calculations explicitly for the five nuclei under study, we obtain the results given in the table below.

Nucleus	$A^{\frac{2}{3}}$	$\frac{Z^2}{A}$	S (MeV)	C (MeV)	$S+C$ (MeV)
$^{228}\mathrm{Th}^{90}$	37.33	35.53	627.1	954.9	1582.0
$^{234}\mathrm{U}^{92}$	37.94	36.17	637.4	988.0	1625.4
$^{236}\mathrm{U}^{92}$	38.19	35.86	641.6	986.0	1627.6
$^{240}\mathrm{Pu}^{94}$	38.56	36.82	647.8	1022.2	1670.0
$^{243}\mathrm{Pu}^{94}$	38.94	36.36	654.2	1019.4	1673.6

Since the surface and the Coulomb energies destabilize a nucleus, and the sum of the two energies is maximum for $^{243}\mathrm{Pu}^{94}$, on the basis of this simple model, we conclude that $^{243}\mathrm{Pu}^{94}$ would fission most easily. (Of course, we are neglecting quantum effects.) However, if we take the deformed liquid drop model, which is at the heart of the theory of nuclear fission, we note that the stability of a nucleus is determined from Eq. (5.7) of the text

$$
\Delta = \frac{1}{5}\epsilon^2\left(2a_2 - a_3\frac{Z^2}{A}\right) \approx \frac{1}{5}\epsilon^2\left(33.6 - 0.72 \times \frac{Z^2}{A}\right) \tag{5.41}
$$

and is controlled by the quantity within the parentheses. The lower this value, the more unstable is the nucleus and, in particular, if it is negative, the nucleus will fission. Tabulating these values for the five nuclei, we obtain

Nucleus	$\left(33.6 - 0.72 \times \frac{Z^2}{A}\right)$ (MeV)
$^{228}\mathrm{Th}^{90}$	8.02
$^{234}\mathrm{U}^{92}$	7.56
$^{236}\mathrm{U}^{92}$	7.78
$^{240}\mathrm{Pu}^{94}$	7.09
$^{243}\mathrm{Pu}^{94}$	7.42

From the values in this table, we conclude that $^{240}\mathrm{Pu}^{94}$ is most likely to fission, while the spherical liquid-drop model suggests $^{243}\mathrm{Pu}^{94}$ as the most likely candidate to fission.

Problem 5.7 *If the efficiency for conversion of heat to electricity is only* 5%, *calculate the rate of consumption of* ^{235}*U fuel in a nuclear reactor operating at a power level of* 500 MW *of electricity.*

As we saw in Problem 5.2 (see Eq. (5.10)), the complete fission of 1 g of ^{235}U yields an energy

$$\begin{aligned}
E_{\text{fission}} &\approx 4.55 \times 10^{23}\,\text{MeV/g} \\
&\approx 4.55 \times 10^{23}\,\text{MeV/g} \times \frac{1}{6.2 \times 10^{5}\,\text{MeV/erg}} \\
&\approx 7.3 \times 10^{17}\,\text{erg/g} = 7.3 \times 10^{10}\,\text{Joules/g} \\
&= 7.3 \times 10^{10}\,(\text{Joules/sec/g}) \times 1\,\text{sec} \\
&= 7.3 \times 10^{10}\,\text{Watt/g} \times \frac{1\,\text{sec}}{8.64 \times 10^{4}\,\text{sec/day}} \\
&\approx 8.7 \times 10^{5}\,\text{WD/g} = 0.87\,\text{MWD/g}, \qquad (5.42)
\end{aligned}$$

where we have used the conversion from "MeV" to "erg" given in Eq. (2.2). We note that since the efficiency of conversion of heat to electricity is only 5%, the fission of 1 g of ^{235}U will yield less useful electric energy:

$$E_{\text{fission}}^{(\text{electricity})} = 0.05 \times 0.87\,\text{MWD/g} \approx 0.043\,\text{MWD/g}. \qquad (5.43)$$

Therefore, to produce 500 MWD of electricity, we need

$$\frac{500\,\text{MWD}}{E_{\text{fission}}^{(\text{electricity})}} \approx \frac{500\,\text{MWD}}{0.043\,\text{MWD/g}} \approx 12{,}500\,\text{g} \approx 12.5\,\text{kg} \qquad (5.44)$$

of ^{235}U per day. This represents a large rate of fuel consumption, but many orders of magnitude smaller than for the case of fossil fuel.

Problem 5.8 *In the fission of* ^{235}U, *the mass ratio of the two produced fission fragments is* 1.5. *What is the ratio of the velocities of these fragments?*

The fission of ^{235}U can be represented as a two-body decay, e.g.,

$$^{235}\text{U} \rightarrow {}^{148}\text{La} + {}^{87}\text{Br}. \tag{5.45}$$

As we saw in Problem 5.2 (see Eq. (5.8)), a single fission yields an energy of

$$Q \approx 178\,\text{MeV}. \tag{5.46}$$

This implies that the fission fragments are nonrelativistic. If we assume the two fragments have masses M_1 and M_2 with

$$\frac{M_1}{M_2} = 1.5, \tag{5.47}$$

and if we further assume that ^{235}U decays from rest, then from momentum conservation we can write that $M_1 v_1 = M_2 v_2$, and obtain

$$\frac{v_1}{v_2} = \frac{M_2}{M_1} = \frac{1}{1.5} \approx 0.67. \tag{5.48}$$

This represents the ratio of the velocities (magnitudes) of the two fission fragments.

Problem 5.9 *How much energy is liberated when* 1 gram *of hydrogen atoms is converted into helium atoms through fusion? Compare this with the energy liberated in the fission of* 1 gm *of* ^{235}U.

From Eq. (5.17) of the text, we get

$$4(^1\text{H}) \rightarrow {}^4\text{He} + 2e^+ + 2\nu_e + 2\gamma + 24.68\,\text{MeV}. \tag{5.49}$$

Namely, four nuclei of hydrogen, through fusion, lead to a helium nucleus with a release of 24.68 MeV of energy. In 1 g of hydrogen, we have

$$N_{^1\mathrm{H}} \approx \frac{6 \times 10^{23}}{1} = 6 \times 10^{23} \tag{5.50}$$

nuclei of hydrogen. Therefore, the complete fusion of 1 g of hydrogen will yield

$$\begin{aligned} E_{\text{fusion}} &= \frac{N_{^1\mathrm{H}}}{4} \times 24.68\,\mathrm{MeV} \\ &\approx \frac{6 \times 10^{23}}{4} \times 24.68\,\mathrm{MeV} \approx 3.7 \times 10^{24}\,\mathrm{MeV} \end{aligned} \tag{5.51}$$

of energy. Comparing this with the amount of energy released in the complete fission of 1 g of ^{235}U (derived in Eq. (5.10)),

$$E_{\text{fission}} \approx 4.55 \times 10^{23}\,\mathrm{MeV}, \tag{5.52}$$

we conclude that the fusion of 1 g of hydrogen releases approximately 8 times the energy released in the complete fission of 1 g of ^{235}U.

Problem 5.10 *The half life of radioactive cobalt-*60 *is* 5.26 yr.

(a) *Calculate its mean life and disintegration constant.*
(b) *What is the activity of* 1 gm *of* ^{60}Co*? Express this in curies and in rutherfords.*
(c) *What is the mass of a* 10-Ci *sample of cobalt-*60*?*

We are given that ^{60}Co has a half-life

$$t_{1/2}^{(^{60}\mathrm{Co})} = 5.26\,\mathrm{yr} \approx 5.26 \times 3.1 \times 10^7\,\mathrm{sec} \approx 1.6 \times 10^8\,\mathrm{sec}. \tag{5.53}$$

(a) It follows from the definitions that

$$\begin{aligned} \tau^{(^{60}\mathrm{Co})} &= \text{mean life} = \frac{t_{1/2}^{(^{60}\mathrm{Co})}}{\ln 2} \approx \frac{1.6 \times 10^8\,\mathrm{sec}}{0.693} \approx 2.3 \times 10^8\,\mathrm{sec}, \\ \lambda^{(^{60}\mathrm{Co})} &= \text{decay constant} = \frac{1}{\tau^{(^{60}\mathrm{Co})}} \approx \frac{1}{2.3 \times 10^8\,\mathrm{sec}} \\ &\approx 4.3 \times 10^{-9}/\mathrm{sec}. \end{aligned} \tag{5.54}$$

(b) One gram of ^{60}Co has

$$N_{^{60}\mathrm{Co}} \approx \frac{6 \times 10^{23}}{60} = 10^{22} \tag{5.55}$$

nuclei of ^{60}Co. The activity of 1 g of ^{60}Co is therefore

$$\begin{aligned}\mathcal{A}(0) &= \lambda^{(^{60}\mathrm{Co})} N_{^{60}\mathrm{Co}} \approx 4.3 \times 10^{-9}/\mathrm{sec} \times 10^{22}\,\mathrm{decays} \\ &= 4.3 \times 10^{13}\,\mathrm{decays/sec}. \end{aligned} \tag{5.56}$$

(c) The sample with 10 Ci activity has

$$10\,\mathrm{Ci} = 10 \times 3.7 \times 10^{10}\,\mathrm{decays/sec} = 3.7 \times 10^{11}\,\mathrm{decays/sec}. \tag{5.57}$$

Since the activity of 1 g of ^{60}Co is 4.3×10^{13} decays/sec, we conclude that the sample must have a mass

$$\frac{3.7 \times 10^{11}\,\mathrm{decays/sec}}{4.3 \times 10^{13}\,\mathrm{decays/sec/g}} \approx 0.86 \times 10^{-2}\,\mathrm{g} = 8.6\,\mathrm{mg}. \tag{5.58}$$

Problem 5.11 *Suppose that atoms of type* 1 *decay to type* 2*, which, in turn, decay to stable atoms of type* 3*. The decay constants of* 1 *and* 2 *are* λ_1 *and* λ_2*, respectively. Assume that at* $t = 0$, $N_1 = N_0$ *and* $N_2 = N_3 = 0$. *What are the values for* $N_1(t)$, $N_2(t)$ and $N_3(t)$ *at any later time* t*?*

When several species of radioactive material are involved in a decay sequence

$$1 \to 2 \to 3 \to \cdots, \tag{5.59}$$

the dynamical equation for the number of species at any time is given in Eq. (5.31) of the text. Extending this to the present problem of a decay involving three species

$$1 \to 2 \to 3, \tag{5.60}$$

where species "3" is stable ($\lambda_3 = 0$), the relationships between species become

$$\begin{aligned}\frac{\mathrm{d}N_1(t)}{\mathrm{d}t} &= -\lambda_1 N_1(t), \\ \frac{\mathrm{d}N_2(t)}{\mathrm{d}t} &= \lambda_1 N_1(t) - \lambda_2 N_2(t), \\ \frac{\mathrm{d}N_3(t)}{\mathrm{d}t} &= \lambda_2 N_2(t), \end{aligned} \tag{5.61}$$

where λ_1 and λ_2 represent the decay constants of the first two unstable species.

We can solve the equations in Eq. (5.61) subject to the initial conditions

$$N_1(0) = N_0, \quad N_2(0) = N_3(0) = 0. \tag{5.62}$$

The first equation in Eq. (5.61) leads to

$$\begin{aligned} \frac{\mathrm{d}N_1(t)}{\mathrm{d}t} &= -\lambda_1 N_1(t) \\ \text{or} \quad \int_{N_0}^{N_1(t)} \frac{\mathrm{d}N_1}{N_1} &= -\lambda_1 \int_0^t \mathrm{d}t \\ \text{or} \quad \ln \frac{N_1(t)}{N_0} &= -\lambda_1 t \\ \text{or} \quad N_1(t) &= N_0 e^{-\lambda_1 t}, \end{aligned} \tag{5.63}$$

where the initial conditions appear in the limits of integrations.

Substituting this into the second equation of Eq. (5.61), we obtain

$$\frac{\mathrm{d}N_2(t)}{\mathrm{d}t} = \lambda_1 N_0 e^{-\lambda_1 t} - \lambda_2 N_2(t). \tag{5.64}$$

Multiplying both sides with $e^{\lambda_2 t}$ (integrating factor), we obtain

$$\begin{aligned} e^{\lambda_2 t}\left(\frac{\mathrm{d}N_2(t)}{\mathrm{d}t} + \lambda_2 N_2(t)\right) &= \lambda_1 N_0 e^{-(\lambda_1-\lambda_2)t} \\ \text{or} \quad \int_0^t \mathrm{d}t \, \frac{\mathrm{d}\left(e^{\lambda_2 t} N_2(t)\right)}{\mathrm{d}t} &= \int_0^t \mathrm{d}t \, \lambda_1 N_0 \, e^{-(\lambda_1-\lambda_2)t} \\ \text{or} \quad e^{\lambda_2 t} N_2(t) &= -\frac{\lambda_1 N_0}{\lambda_1 - \lambda_2}\left(e^{-(\lambda_1-\lambda_2)t} - 1\right) \\ \text{or} \quad N_2(t) &= \frac{\lambda_1 N_0}{\lambda_1 - \lambda_2}\left(e^{-\lambda_2 t} - e^{-\lambda_1 t}\right). \end{aligned} \tag{5.65}$$

In evaluating the integral, we used the initial conditions, and the solution therefore automatically satisfies Eq. (5.62).

Finally, substituting this solution into the last of equations in Eq. (5.61) we have

$$\begin{aligned}\frac{\mathrm{d}N_3(t)}{\mathrm{d}t} &= \lambda_2 N_2(t) = \frac{\lambda_1\lambda_2 N_0}{\lambda_1-\lambda_2}(e^{-\lambda_2 t}-e^{-\lambda_1 t})\\ \text{or}\quad N_3(t) &= \frac{\lambda_1\lambda_2}{\lambda_1-\lambda_2}\left(-\frac{1}{\lambda_2}(e^{-\lambda_2 t}-1)+\frac{1}{\lambda_1}(e^{-\lambda_1 t}-1)\right)\\ &= \frac{N_0}{\lambda_1-\lambda_2}(\lambda_2 e^{-\lambda_1 t}-\lambda_1 e^{-\lambda_2 t}+(\lambda_1-\lambda_2)),\end{aligned} \tag{5.66}$$

where we have again used the initial conditions in evaluating the integrals, which automatically satisfy Eq. (5.62).

Thus, the complete solution satisfying the initial conditions of Eq. (5.62) takes the form

$$\begin{aligned}N_1(t) &= N_0 e^{-\lambda_1 t},\\ N_2(t) &= \frac{\lambda_1 N_0}{\lambda_1-\lambda_2}(e^{-\lambda_2 t}-e^{-\lambda_1 t}),\\ N_3(t) &= \frac{N_0}{\lambda_1-\lambda_2}(\lambda_2 e^{-\lambda_1 t}-\lambda_1 e^{-\lambda_2 t}+(\lambda_1-\lambda_2)).\end{aligned} \tag{5.67}$$

These satisfy the given initial conditions, and as $t \to \infty$ we obtain

$$N_1(t) \to 0, \quad N_2(t) \to 0, \quad N_3(t) \to N_0, \tag{5.68}$$

signifying that all the initial particles decay eventually into the stable variety "3".

Problem 5.12 *The activity of a certain material decreases by a factor of* 8 *in a time interval of* 30 *days. What is its half life, mean life and disintegration constant?*

The activity of a radioactive material has an exponential time dependence given in Eq. (5.26) of the text

$$\begin{aligned}\mathcal{A}(t) &= \mathcal{A}(0)e^{-\lambda t}\\ \text{or}\quad \ln\left(\tfrac{\mathcal{A}(t)}{\mathcal{A}(0)}\right) &= -\lambda t.\end{aligned} \tag{5.69}$$

If the activity of a material decreases by a factor of 8 in 30 days, we have

$$t = 30\,\text{days} \approx 30 \times 8.6 \times 10^4\,\text{sec} = 2.6 \times 10^6\,\text{sec},$$
$$\frac{\mathcal{A}(t = 30\,\text{days})}{\mathcal{A}(0)} = \frac{1}{8}. \tag{5.70}$$

Substituting this into Eq. (5.69) we obtain

$$\lambda \times 2.6 \times 10^6\,\text{sec} \approx -\ln\frac{1}{8}$$
$$\text{or} \quad \lambda = -\frac{1}{2.6 \times 10^6\,\text{sec}} \times (-2.08) \approx 8 \times 10^{-7}/\text{sec}. \tag{5.71}$$

From this value of the decay constant (disintegration constant), we can determine

$$\tau = \text{mean life} = \frac{1}{\lambda} \approx \frac{1}{8 \times 10^{-7}/\text{sec}} = 1.25 \times 10^6\,\text{sec}$$
$$\approx 1.25 \times 10^6 \times \frac{1\,\text{sec}}{8.6 \times 10^4\,\text{sec/day}} \approx 14.5\,\text{days}, \tag{5.72}$$
$$t_{1/2} = \tau \ln 2 \approx 1.25 \times 10^6\,\text{sec} \times 0.693 \approx 8.7 \times 10^5\,\text{sec}$$
$$\approx 8.7 \times 10^5 \times \frac{1\,\text{sec}}{8.6 \times 10^4\,\text{sec/day}} \approx 10.1\,\text{days}.$$

Problem 5.13 *For a prolate spheroid (ellipsoid) with eccentricity x, the semi-major axis a and semi-minor axis b in Fig. 5.2 are related through $b = \sqrt{1-x^2}\,a$. If the volume and surface area of the nuclear ellipsoid are given, respectively, as $\frac{4}{3}\pi ab^2$ and $2\pi b\left(b + \frac{a \sin^{-1} x}{x}\right)$, defining $\epsilon = \frac{1}{3}x^2$, show that Eq. (5.5) holds for small values of x. (Hint: Assume that the volume does not change under distortion; expand functions of x, and keep all terms up to order x^5.) Using this result, roughly, how would you argue that Eq. (5.6) has the right dependence?*

The ellipsoid has a semi-major axis a and a semi-minor axis b, related as

$$b = \sqrt{1 - x^2}\; a, \tag{5.73}$$

where x represents the eccentricity of the ellipsoid. If we assume that the volume of the ellipsoid is the same as that of a sphere with radius R, then we have

$$V = \frac{4}{3}\,\pi R^3 = \frac{4}{3}\,\pi a b^2. \tag{5.74}$$

As pointed out in Eq. (5.3) of the text, this leads to the relationship

$$a = R(1+\epsilon), \quad b = \frac{R}{(1+\epsilon)^{\frac{1}{2}}}. \tag{5.75}$$

Using this in Eq. (5.73), we obtain

$$\begin{aligned} \frac{R}{(1+\epsilon)^{\frac{1}{2}}} &= (1-x^2)^{\frac{1}{2}} R(1+\epsilon) \\ \text{or} \quad 1+\epsilon &= (1-x^2)^{-\frac{1}{3}}. \end{aligned} \tag{5.76}$$

We see that for small x, we can Taylor expand this to obtain

$$\epsilon \approx -1 + \left(1 + \frac{1}{3}x^2\right) = \frac{1}{3}x^2, \tag{5.77}$$

as we expect. In general, using Eqs. (5.75) and (5.76) we can write

$$a = R(1-x^2)^{-\frac{1}{3}}, \quad b = R(1-x^2)^{\frac{1}{6}}. \tag{5.78}$$

The surface area of the ellipsoid can be written as

$$\begin{aligned} S &= 2\pi b\left[b + a \times \frac{\sin^{-1} x}{x}\right] \\ &= 2\pi R(1-x^2)^{\frac{1}{6}}\left[R(1-x^2)^{\frac{1}{6}} + R(1-x^2)^{-\frac{1}{3}} \times \frac{\sin^{-1} x}{x}\right] \\ &= 2\pi R^2\left[(1-x^2)^{\frac{1}{3}} + (1-x^2)^{-\frac{1}{6}} \times \frac{\sin^{-1} x}{x}\right]. \end{aligned} \tag{5.79}$$

For small x, we can Taylor expand the above quantities to obtain

$$\begin{aligned} (1-x^2)^{\frac{1}{3}} &\approx 1 - \frac{1}{3}x^2 - \frac{1}{9}x^4, \\ (1-x^2)^{-\frac{1}{6}} &\approx 1 + \frac{1}{6}x^2 + \frac{7}{72}x^4, \\ \frac{\sin^{-1} x}{x} &\approx \frac{1}{x}\left(x + \frac{1}{6}x^3 + \frac{3}{40}x^5\right) = 1 + \frac{1}{6}x^2 + \frac{3}{40}x^4. \end{aligned} \tag{5.80}$$

Substituting this into Eq. (5.79), we obtain the surface area for small x (keeping terms up to x^4):

$$\begin{aligned}
S &\approx 2\pi R^2 \left[1 - \frac{1}{3}x^2 - \frac{1}{9}x^4 + \left(1 + \frac{1}{6}x^2 + \frac{7}{72}x^4\right)\left(1 + \frac{1}{6}x^2 + \frac{3}{40}x^4\right)\right] \\
&= 2\pi R^2 \left[1 - \frac{1}{3}x^2 - \frac{1}{9}x^4 + 1 + \frac{1}{6}x^2 + \frac{3}{40}x^4 + \frac{1}{6}x^2 + \frac{1}{36}x^4 + \frac{7}{72}x^4\right] \\
&= 4\pi R^2 \left(1 + \frac{2}{45}x^4\right) = 4\pi R^2 \left(1 + \frac{2}{5}\left(\frac{x^2}{3}\right)^2\right) \\
&= 4\pi R^2 \left(1 + \frac{2}{5}\epsilon^2\right).
\end{aligned} \tag{5.81}$$

This shows how the surface of the sphere scales under the volume-preserving deformation. The change can be described by the scaling behavior

$$\begin{aligned}
R^2 &\rightarrow R^2 \left(1 + \frac{2}{5}\epsilon^2\right) \\
\text{or} \quad A^{\frac{2}{3}} &\rightarrow A^{\frac{2}{3}} \left(1 + \frac{2}{5}\epsilon^2\right),
\end{aligned} \tag{5.82}$$

where we have used the fact that for nuclei

$$R \approx A^{\frac{1}{3}}. \tag{5.83}$$

The scaling in Eq. (5.83) reproduces Eq. (5.5) of the text. We can also write:

$$A^{-\frac{1}{3}} \rightarrow A^{-\frac{1}{3}} \left(1 + \frac{2}{5}\epsilon^2\right)^{-\frac{1}{2}} \approx A^{-\frac{1}{3}} \left(1 - \frac{1}{5}\epsilon^2\right), \tag{5.84}$$

which leads to Eq. (5.6) of the text.

Problem 5.14 *Secular equilibrium can also be defined through the requirement that*

$$\frac{d}{dt}\left(\frac{N_2}{N_1}\right) = \frac{d}{dt}\left(\frac{N_3}{N_2}\right) = \frac{d}{dt}\left(\frac{N_4}{N_3}\right) \cdots = 0.$$

Assuming $\lambda_1 \ll \lambda_2, \lambda_3, \lambda_4, \ldots$, *show explicitly that you retrieve the first three relations in Eq.* (5.33). *What happens for the final state of the decay chain? Is this sensible?*

Let us consider a sequence of radioactive decays

$$1 \to 2 \to 3 \to \cdots \to n, \tag{5.85}$$

described by the relationships

$$\begin{aligned} \frac{\mathrm{d}N_1}{\mathrm{d}t} &= -\lambda_1 N_1, \\ \frac{\mathrm{d}N_2}{\mathrm{d}t} &= \lambda_1 N_1 - \lambda_2 N_2, \\ \frac{\mathrm{d}N_3}{\mathrm{d}t} &= \lambda_2 N_2 - \lambda_3 N_3, \\ &\vdots \\ \frac{\mathrm{d}N_n}{\mathrm{d}t} &= \lambda_{n-1} N_{n-1}, \end{aligned} \tag{5.86}$$

with the condition on the decay constants that $\lambda_1 \ll \lambda_2, \lambda_3, \ldots$.

The condition for secular equilibrium

$$\frac{\mathrm{d}\left(\frac{N_\alpha}{N_{\alpha+1}}\right)}{\mathrm{d}t} = 0, \quad \alpha = 1, 2, 3, \ldots, \tag{5.87}$$

can also be written as

$$\frac{\mathrm{d}\left(\frac{N_1}{N_{\alpha+1}}\right)}{\mathrm{d}t} = 0. \tag{5.88}$$

Using Eq. (5.86), we note that

$$\begin{aligned} \frac{\mathrm{d}\left(\frac{N_1}{N_{i+1}}\right)}{\mathrm{d}t} &= \frac{1}{N_{i+1}}\frac{\mathrm{d}N_1}{\mathrm{d}t} - \frac{N_1}{(N_{i+1})^2}\frac{\mathrm{d}N_{i+1}}{\mathrm{d}t} \\ &= \frac{1}{N_{i+1}}\left(-\lambda_1 N_1 - \frac{N_1}{N_{i+1}}\left(\lambda_i N_i - \lambda_{i+1} N_{i+1}\right)\right) \\ &= \frac{N_1}{(N_{i+1})^2}((\lambda_{i+1} - \lambda_1) N_{i+1} - \lambda_i N_i) \\ &\approx \frac{N_1}{(N_{i+1})^2}\left(\lambda_{i+1} N_{i+1} - \lambda_i N_i\right), \end{aligned} \tag{5.89}$$

where we are assuming that $i = 1, 2, \ldots, n-2$, i.e. this holds for all but the last species. We have also used the fact that $\lambda_1 \ll \lambda_{i+1}$. At secular equilibrium, we expect Eq. (5.89) to vanish, leading to the relationship

$$\lambda_{i+1}N_{i+1} - \lambda_i N_i = 0, \quad i = 1, 2, \ldots, n-2. \tag{5.90}$$

This can be written out explicitly as

$$\lambda_1 N_1 = \lambda_2 N_2 = \cdots = \lambda_{n-1}N_{n-1}, \tag{5.91}$$

and reflects the fact that, for secular equilibrium, there are as many particles of a given species produced as there are decaying (not counting particles of the final variety).

To analyze the behavior of the final variety of particles, we note from Eq. (5.86) that

$$\begin{aligned}
\frac{\mathrm{d}\left(\frac{N_1}{N_n}\right)}{\mathrm{d}t} &= \frac{1}{N_n}\frac{\mathrm{d}N_1}{\mathrm{d}t} - \frac{N_1}{(N_n)^2}\frac{\mathrm{d}N_n}{\mathrm{d}t} \\
&= \frac{1}{N_n}(-\lambda_1 N_1) - \frac{N_1}{(N_n)^2}(\lambda_{n-1}N_{n-1}) \\
&= -\frac{N_1}{(N_n)^2}(\lambda_1 N_n + \lambda_1 N_1) \\
&= -\lambda_1\left[\left(\frac{N_1}{N_n}\right) + \left(\frac{N_1}{N_n}\right)^2\right] \neq 0,
\end{aligned} \tag{5.92}$$

where we have used the relationships in Eq. (5.91). If we assume that λ_1 is very small, such that $\lambda_1 N_1 \approx 0$ (which will guarantee that in secular equilibrium N_1 does not change appreciably), the right-hand side of the above equation will not vanish because $\lambda_1(N_1)^2$ need not be negligible. The nonconstant nature of $\frac{N_1}{N_n}$ simply reflects the fact that even though N_1 may be constant in secular equilibrium, N_n cannot be because there is production of particles of the final variety, but no decay.

6. Deposition of Energy in Media

Problem 6.1 *What is the minimum thickness of aluminum in cm that is needed to stop a* 3 MeV α *particle? What about the thickness needed to stop a* 3 MeV *electron? (Use the approximate range-energy relationship provided in Examples* 1 *and* 2.)

Using the approximate range-energy relationships given in Examples 6.1 and 6.2, the range of an α particle with kinetic energy T (MeV) in air and aluminum is

$$\begin{aligned} \text{Air} : R_{\text{air}} &= 0.32T^{3/2} = 0.32(3)^{1.5} = 1.7\,\text{cm}, \\ \text{Al} : R &= R_{\text{air}}/1600 = (1.7)/1600 = 10\ \mu\text{m}. \end{aligned} \tag{6.1}$$

For an electron, the range $R\left(\text{g}\cdot\text{cm}^{-2}\right)$ is

$$R = 0.53T\,(\text{MeV}) - 0.16 = 0.53(3) - 0.16 = 1.43\,\text{g}\cdot\text{cm}^{-2} \tag{6.2}$$

and dividing by the density of aluminum $\rho = 2.7\,\text{g/cm}^3$, yields the range in cm,

$$R = 1.43\,\text{g}\cdot\text{cm}^{-2}/(2.7\,\text{g/cm}^3) = 0.53\,\text{cm}. \tag{6.3}$$

Problem 6.2 *About how much steel in cm is required to stop a* 500 GeV *muon if the muon deposits energy only via ionization loss? (Use Eq.* (6.5) *to calculate your result.) Would you need a comparable amount of material to stop* 500 GeV *electrons? What about* 500 GeV *protons?*

Muons: A 500 GeV muon will be a minimum-ionizing particle for most of its path through steel. If we assume that steel has the same density and atomic number as iron, $\rho_{\text{Fe}} = 7.9\,\text{g/cm}^3$, and $A_{\text{Fe}} = 56$, respectively, the muon will lose energy by ionization at a rate given by

$$S_{\text{min}} \approx 3.5\rho\frac{Z}{A}\,\text{MeV/cm} = 3.5(7.9)\left(\frac{26}{56}\right) = 12.8\,\text{MeV/cm} \qquad (6.4)$$

which yields an approximate range of

$$R_\mu = \frac{E}{S} = \frac{500\times10^3\,\text{MeV}}{12.8\,\text{MeV/cm}} = 390\,\text{m}. \qquad (6.5)$$

Electrons: For a 500 GeV electron in steel, the primary energy loss is via bremsstrahlung, with a radiation length of $X_0 = 1.76$ cm. High-energy bremsstrahlung photons will convert to e^+e^- pairs in a characteristic length of $X_{\text{pair}} = 9X_0/7$. The electron and positron of the pair will then radiate bremsstrahlung photons. This cycle repeats many times resulting in an "electromagnetic shower" of electrons, positrons and photons. When the average photon energy drops below the pair-production threshold, the shower terminates quickly. The depth (in units of X_0) of the maximum particle density, t_{max}, increases logarithmically with energy, and can be parametrized by a scale factor known as the "critical energy" E_c. $E_c = 23$ MeV for iron, which yields

$$t_{\text{max}}(E) \approx \ln(E/E_c) \quad \text{with } E \text{ in MeV}. \qquad (6.6)$$

A typical energy-deposition profile for a 30 GeV electron, with a maximum at $t_{\text{max}}(30\,\text{GeV}) = \ln(3\times10^4/23) = 7.2X_0$, is shown in Fig. 6.1. A 30 GeV shower is contained within $\approx 20X_0$ (≈ 0.35 m). The depth of the point of maximum energy deposition for 500 GeV electrons is

$$t_{\text{max}}(500\,\text{GeV}) = \ln(5\times10^5/23) = 10X_0.$$

Assuming that shower containment scales as t_{max}, a 500 GeV electron shower is contained within $\approx 28X_0$.

Protons: Protons interact with matter via the strong interaction that can be characterized by an interaction length L_{int}, which for iron is $L_{\text{int}}(\text{Fe}) \approx 17$ cm, as given in the Particle Data Group (PDG) tables. Assuming that a nucleon loses on average 1/2 of its energy in each interaction, the number of interactions, n, needed to drop the initial energy below 1 GeV, is defined by $(500\,\text{GeV})(0.5)^n = 1\,\text{GeV}$, or $n = \ln 500/\ln 2 \approx 9$, or $(9\times17\,\text{cm}) \approx 1.6$ m of iron. Assuming that

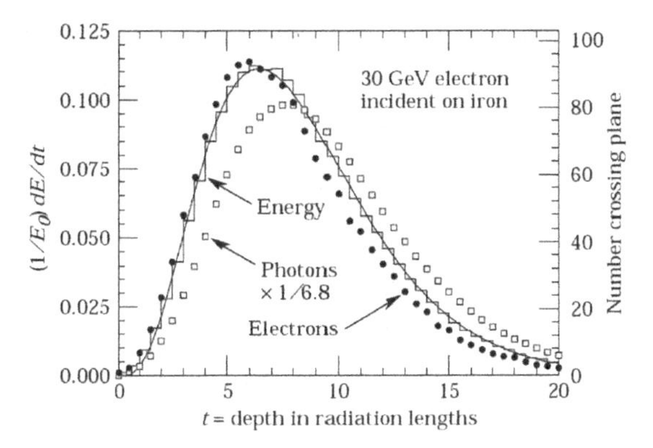

Fig. 6.1. Energy deposition profile for a 30 GeV electron in Iron, along with the number of electrons and photons in the shower, as a function of depth in units of radiation lengths.

a $T = 1\,\text{GeV}$ proton ($pc \approx 1.7\,\text{GeV}$) loses energy through ionization down to the level of the critical energy, then, for $I_{\text{min}} \approx 12\,\text{MeV/cm}$ in Fe, this requires another 0.8 m of iron, and corresponds to a lower limit on the length needed for the proton to dissipate its final 1 GeV of energy. Hence, a total length of ≈ 2.4 m of iron is required for the absorption of the energy of an incident 500 GeV proton.

Problem 6.3 *Multiple-scattering error often limits the ability to measure the direction of motion of a charged particle. To what accuracy can the incident angle of a* 500 GeV *muon be measured after the particle traverses one meter of iron?*

A particle with a momentum p travelling through a medium of length L and radiation length X_0 will multiple scatter through an rms angle θ_{rms} given by

$$\theta_{\text{rms}} = \frac{20\,\text{MeV}}{\beta pc}\sqrt{\frac{L}{X_0}}. \tag{6.7}$$

A 500 GeV muon ($pc = 500\,\text{GeV}$, $\beta \approx 1$), traversing a length $L = 100\,\text{cm}$ of Fe (radiation length $X_0 = 1.8\,\text{cm}$) will multiple scatter through an rms angle θ_{rms} relative to the incident direction of the muon:

$$\theta_{\text{rms}} = \frac{20\,\text{MeV}}{\beta pc}\sqrt{\frac{L}{X_0}} = \frac{20}{5 \times 10^5}\sqrt{\frac{100}{1.8}} \approx 0.3\,\text{mrad}. \tag{6.8}$$

Problem 6.4 *Typically, what fraction of a beam of* 100 GeV *photons will be transmitted through a* 2 cm *thick lead absorber?*

The fraction of high-energy photons transmitted through a medium of radiation length X_0 and length x is

$$I/I_0 = e^{-\mu x}, \quad \text{where } \mu = \frac{7}{9X_0}. \tag{6.9}$$

For Pb, $X_0 = 0.56$ cm, and for a thickness of 2 cm, the fraction of photons transmitted is

$$I/I_0 = e^{-(1.4)2} = 6.1\%. \tag{6.10}$$

Problem 6.5 *The capture cross section for thermal neutrons on* ^{27}*Al is* 233 mb. *On average, how far can a beam of such neutrons penetrate a slab of aluminum* (ρ = 2.7 gm/cm^3) *before half of the beam is absorbed. (See relation* (6.27).)

The cross section of thermal neutrons in Al is 233 mb ($\sigma = 2.33 \times 10^{-25}$ cm^2). A beam of neutrons travelling through a thickness x of aluminum, with $\rho = 2.7$ g/cm^3, attenuates as $I = I_0 e^{-\mu x}$, with an attenuation constant μ:

$$\mu = \rho \frac{A_0}{A} \sigma = 2.7 \frac{6 \times 10^{23}}{27} 2.3 \times 10^{-25} = 1.4 \times 10^{-2}\,\text{cm}^{-1}. \tag{6.11}$$

To attenuate neutrons by a factor of two means that:

$$-\mu x = \ln(I/I_0) = \ln(0.5) = -0.693,$$

and therefore

$$x = 0.693/1.4 \times 10^{-2}\,\text{cm}^{-1} \approx 50\,\text{cm}. \tag{6.12}$$

Problem 6.6 *Protons and* α*-particles of* 20 MeV *pass through* 0.001 cm *of aluminum foil. How much energy do such particles deposit within the foil?*

For nonrelativistic particles, the stopping power is given by Eq. (6.3) in the text:

$$S(T) = \frac{4\pi Q^2 e^2 n Z}{m_e c^2 \beta^2} \ln\left[\frac{2m_e\beta^2c^2}{\bar{I}}\right] = K z_Z^2 \frac{Z}{A}\frac{1}{\beta^2} \ln\left[\frac{2m_e\beta^2c^2}{\bar{I}}\right];$$

$$K = 0.31\,\text{MeV}\cdot\text{g}^{-1}\cdot\text{cm}^2.$$

To obtain the velocity (β) of a particle with mass m and kinetic energy T, we use the relationship:

$$\beta = \frac{pc}{E} = \frac{\sqrt{2mT}}{m} = \sqrt{\frac{2T}{m}}. \tag{6.13}$$

For protons and α particles of kinetic energy 20 MeV passing through 0.001 cm of Al foil, the parameters for stopping power are:

$$\beta_p^2 = \frac{40}{938} = 0.043; \quad \beta_\alpha^2 = \frac{40}{3730} = 0.011;$$
$$m_p = 938\,\text{MeV}; \quad m_\alpha = 3730\,\text{MeV}; \quad z_p = 1; \quad z_\alpha = 2;$$
$$Z_{\text{Al}} = 13; \quad A_{\text{Al}} = 27; \quad \rho_{\text{Al}} = 2.7\,\text{g}\cdot\text{cm}^{-3}; \quad \bar{I}_{\text{Al}} \approx 27\,\text{eV}.$$

The stopping power and energy deposition for protons in 0.001 cm of Al is:

$$S(T)_p = (0.31)\frac{13}{27}\frac{1}{0.043}\ln\left[\frac{1.02\times10^6(0.043)}{130}\right]$$
$$= (3.5)(5.8) = 20\,\text{MeV}\cdot\text{cm}^2/\text{g}, \tag{6.14}$$
$$\rho S(T)_p dx = (2.7)(20)(0.001)\,\text{MeV} = 54\,\text{keV}$$

while for α particles, we obtain:

$$S(T)_\alpha = (0.31)\frac{13}{27}\frac{1}{0.011}\ln\left[\frac{1.02\times10^6(0.011)}{130}\right]$$
$$= (13.6)(4.46) = 61\,\text{MeV}\cdot\text{cm}^2/\text{g}, \tag{6.15}$$
$$\rho S(T)_\alpha dx = (2.7)(61)(0.001)\,\text{MeV} = 160\,\text{keV}.$$

Problem 6.7 *Compare the stopping power of electrons, protons and α-particles in copper, for particle velocities of* $0.5c$.

Stopping powers of electrons, protons and α particles for $v = 0.5c$ can be compared by noting that the stopping power is proportional to Q^2/v^2. Therefore, the stopping powers of electrons and protons, both with $Q = e$, are the same at the same *velocity*. Compared to objects with charge $Q = 1e$, the stopping power for α particles, $Q = 2e$, would therefore be a factor of 4 larger.

Problem 6.8 *Calculate the mass (i.e. $\sqrt{s}$) of the virtual electron in Fig. 6.4 for an incident photon of wavelength of* 1.25×10^{-10} cm. *What is the approximate lifetime of such an object? Repeat your calculation for a wavelength of* 1.25×10^{-12} cm.

To find the mass of the virtual electron for an incident photon with wavelength λ, first find the energy of the incident photon using

$$E_\gamma = hc/\lambda = 2\pi(197\,\text{MeV} \cdot \text{fm})/\lambda. \tag{6.16}$$

For a stationary target electron, the energy and momentum of the virtual electron is,

$$E_v = E_\gamma + m_e c^2, \qquad p_v c = p_\gamma c = E_\gamma. \tag{6.17}$$

The square of the effective mass of this excited electron is

$$m_v^2 c^4 = E_v^2 - p_v^2 c^2 = (E_\gamma + m_e c^2)^2 - E_\gamma^2 = 2E_\gamma m_e c^2 + m_e^2 c^4$$

and

$$m_v = m_e \sqrt{1 + \frac{2E_\gamma}{m_e c^2}}. \tag{6.18}$$

For $\lambda = 1.25 \times 10^{-10}$ cm

$$E_\gamma \approx 1\,\text{MeV} \quad \text{and} \quad m_v = m_e \sqrt{1 + \frac{2(1)}{m_e c^2}} = 2.2 m_e \approx 1\,\text{MeV}/c^2. \tag{6.19}$$

For $\lambda = 1.25 \times 10^{-12}$ cm

$$E_\gamma \approx 100\,\text{MeV} \quad \text{and} \quad m_v = m_e \sqrt{1 + \frac{2(100)}{m_e c^2}} \approx 20 m_e = 10\,\text{MeV}/c^2. \tag{6.20}$$

The Heisenberg uncertainty relationship can be used to estimate the lifetime of the virtual state. For a 1 MeV state

$$\delta t_{\text{cm}} \approx \frac{h/2\pi}{\delta m c^2} = \frac{6.6 \times 10^{-22}\,\text{MeV} \cdot \text{s}}{10\,\text{MeV}} = 6.6 \times 10^{-22}\,\text{s} \tag{6.21}$$

while for a 10 MeV state the lifetime estimate is

$$\delta t_{\text{cm}} \approx \frac{h/2\pi}{\delta mc^2} = \frac{6.6 \times 10^{-22}\,\text{MeV} \cdot \text{s}}{10\,\text{MeV}} = 6.6 \times 10^{-23}\,\text{s}. \tag{6.22}$$

Problem 6.9 *Consider the collision of a photon with a target of mass M that is initially at rest in the laboratory. Show that the minimum laboratory energy that a photon must have to produce an e^+e^- pair is $E_\gamma = 2m_ec^2(1+\frac{m_e}{M})$. (Hint: Equate the expression for s given in Eqs. (1.64) and (1.65).) Thus the threshold for pair production is essentially $2m_ec^2$.*

The minimum γ energy needed for the reaction $\gamma + M \to M + 2m_e$ to proceed can be obtained from the minimum energy required in the center of mass, $\sqrt{s}$. Equating incident and final values of s, we get

$$\begin{aligned} s &= (E_\gamma + Mc^2)^2 - p_\gamma^2c^2 \\ &= (E_\gamma + Mc^2)^2 - E_\gamma^2 = (Mc^2 + 2m_ec^2)^2 \\ \text{or} \quad s/c^2 &= 2ME_\gamma/c^2 + M^2 = M^2 + 4Mm_e + 4m_e^2 \end{aligned} \tag{6.23}$$

$$\text{and} \quad E_\gamma = 2m_ec^2\left(1 + \frac{m_e}{M}\right).$$

Problem 6.10 *What is the mean free path for nuclear collisions of* 10 GeV *protons in liquid hydrogen if the proton-proton total cross section is* 40 mb? *(Assume a liquid hydrogen density of* 0.07 gm/cm³.*)*

To obtain the mean free path, we find the number of atoms per gram of material. Multiplying this by the density, yields the number of scattering atoms per unit volume. Then, multiplying by the cross section, gives the number of interactions per cm of path length.

$$\begin{aligned} \frac{A_0}{A} &= \frac{(6.02 \times 10^{23}\,\text{molecules} \cdot \text{mole}^{-1})(2\,\text{atoms} \cdot \text{molecule}^{-1})}{2(\text{g} \cdot \text{mole}^{-1})} \\ &= 6.02 \times 10^{23}\,\text{atoms} \cdot \text{g}^{-1} \end{aligned}$$

$$\mu = \rho\frac{A_0}{A}\sigma = (0.071)(6 \times 10^{23})(40 \times 10^{-27}) = 1.7 \times 10^{-3}\,\text{cm}^{-1}.$$

The inverse is the mean free path:

$$\lambda = \frac{1}{\mu} = \frac{1}{1.7 \times 10^{-3}}\,\text{cm} = 6 \times 10^2\,\text{cm}. \tag{6.24}$$

Problem 6.11 *Prove the kinematic relation given in Eq.* (6.22).

The Compton scattering formula is derived by applying energy and momentum conservation to the reaction.

Energy conservation:

$$\begin{aligned} E_e &= h(\nu - \nu') + m_e c^2; \\ E_e^2 &= h^2(\nu - \nu')^2 + 2m_e c^2 h(\nu - \nu') + m_e^2 c^4. \end{aligned} \tag{6.25}$$

Momentum conservation:

$$h\nu' \sin\theta = p_e c \sin\theta_e; \quad h(\nu - \nu' \cos\theta) = p_e c \cos\theta_e. \tag{6.26}$$

Squaring the last two equations:

$$h^2\nu'^2 \sin^2\theta = p_e^2 c^2 \sin^2\theta_e; \quad h^2(\nu - \nu' \cos\theta)^2 = p_e^2 c^2 \cos^2\theta_e. \tag{6.27}$$

Adding the above two equations:

$$h^2(\nu - \nu' \cos\theta)^2 + h^2\nu'^2 \sin^2\theta = p_e^2 c^2 = E_e^2 - m_e^2 c^4. \tag{6.28}$$

Expanding the first term in Eq. (6.27) and substituting Eq. (6.25) for E^2:

$$h^2(\nu^2 - 2\nu\nu' \cos\theta) + h^2\nu'^2 = h^2(\nu - \nu')^2 + 2m_e c^2 h(\nu - \nu'). \tag{6.29}$$

Combining the terms:

$$\begin{aligned} h^2\nu\nu'(1 - \cos\theta) &= m_e c^2 h(\nu - \nu'), \\ \frac{h\nu\nu'}{m_e c^2}(1 - \cos\theta) &= \nu - \nu', \\ \nu' \left[1 + \frac{h\nu}{m_e c^2}(1 - \cos\theta)\right] &= \nu. \end{aligned} \tag{6.30}$$

Finally, the solution:

$$\nu' = \frac{\nu}{\left[1 + \frac{h\nu}{m_e c^2}(1 - \cos\theta)\right]}. \tag{6.31}$$

7. Particle Detection

Problem 7.1 *A radioactive source emits α-particles with kinetic energies of* 4 MeV*. What must be the value of an applied magnetic field so that the radius of curvature of the orbit of the α-particle is* 10 cm*? (Does your answer depend on the kind of medium into which the α-particle is emitted?) Do the same calculation for electrons of same kinetic energy.*

This problem is done most easily employing the form of Chap. 8, in Eq. (8.9′), that relates the radius for circular motion of a charged particle with momentum p and the imposed magnetic field B. The z in the formula is the charge of the particle in units of electron charge:

$$R = \frac{p}{0.3zB}; \quad \text{with } R \text{ in } m, \; B \text{ in Tesla}, \; p \text{ in GeV}/c; \; B = \frac{p}{0.3zR}. \tag{7.1}$$

Considering a 4 MeV (kinetic energy) α particle, the nonrelativistic momentum is:

$$p = \sqrt{2mT} = \sqrt{(2(4)\,(4 \times 10^{-3})}\,\text{GeV}/c = 0.18\,\text{GeV}/c$$

and the field needed for a 10 cm radius is

$$B = \frac{0.18}{0.3(2)(0.1)} = 3\,T. \tag{7.2}$$

Considering an electron of 4 MeV (kinetic energy), the momentum is relativistic, and

$$p = \frac{1}{c}\sqrt{E^2 - m_e^2c^4} = \frac{1}{c}\sqrt{(T + m_ec^2)^2 - m_e^2c^4}$$

$$= \frac{1}{c}\sqrt{T^2 + 2m_ec^2T} = \frac{T}{c}\sqrt{1 + \frac{2m_ec^2}{T}},$$

$$p_e = (0.004\,\text{GeV}/c)\sqrt{1 + \frac{0.001}{0.004}} = 0.0045\,\text{GeV}/c,$$

which, for a 10 cm radius, requires the field:

$$B = \frac{p}{0.3zR} = \frac{0.0045}{0.3(1)(0.1)} = 0.15\,T. \tag{7.3}$$

Problem 7.2 *The mass of a K^+ is 494 MeV/c^2 and that of a π^+ is 140 MeV/c^2. If the rms time resolution of each of two scintillation counters that are 2 m apart is 0.2 nsec, calculate to better than 10% accuracy the momentum at which the system will just be able to resolve a π^+ from a K^+ (by one standard deviation). (Hint: See Eq. (7.10).)*

The task is to find the maximum momentum for which a time resolution $\delta t = 0.2$ ns will just resolve a π meson ($mc^2 = 0.14$ GeV) from a K meson ($mc^2 = 0.49$ GeV) for a 2m flight path. Equation (7.18) is useful only in the approximation that the two masses are nearly equal, which is certainly not the case for these two mesons. You must start with Eq. (7.10), or proceed as follows: assuming a velocity $\beta_\pi = 1$, the time of flight for the π meson is $t = L/c = 7$ ns.

The resolution limit is reached when the uncertainty in the K velocity equals the difference in the two velocities, $\delta\beta_K = \Delta\beta = 1 - \beta_K$. The fractional error in the K velocity is equal to the fractional uncertainty in the time resolution:

$$\frac{\delta\beta_K}{\beta_K} = \frac{\delta t}{t} = 0.03.$$

Using an approximation for large momentum, and solving for p,

$$\frac{\delta\beta_K}{\beta_K} = \frac{1}{\beta_K} - 1 = \sqrt{1 + \frac{m_K^2c^2}{p^2}} - 1 \approx \frac{m_K^2c^2}{2p^2} \quad \text{for } p^2 \gg m^2c^2$$

$$\text{and} \quad p = \frac{m_K}{\sqrt{\frac{2\delta\beta_K}{\beta_K}}} = \frac{0.5}{\sqrt{0.06}} = \frac{0.5}{0.25} = 2.0\,\text{GeV}/c. \tag{7.4}$$

Problem 7.3 *What are the Cherenkov angles for electrons and pions of* 1000 MeV/c *for a radiator with* $n = 1.4$? *What will be the ratio of the number of radiated photons for incident electrons and pions?*

The Cherenkov angle is related to the velocity of the particle and index of refraction n of the medium.

$$\cos\theta_c = \frac{1}{\beta n}. \tag{7.5}$$

The Cherenkov angles for electrons and π mesons of 1 GeV/c, passing through radiator of $n = 1.4$, are

$$\begin{aligned} \text{electron} : \theta_c &= \cos^{-1}\left[\frac{1}{(1)1.4}\right] = 44.4^\circ, \\ \text{pion} : \theta_c &= \cos^{-1}\left[\frac{1}{(0.99)1.4}\right] = 43.8^\circ. \end{aligned} \tag{7.6}$$

The number of radiated photons is related to the Cherenkov angle by

$$N \propto \sin^2\theta_c \tag{7.7}$$

so that the ratio of the number of photons emitted by an electron to those for a π meson is

$$\frac{N_e}{N_\pi} = \left[\frac{\sin(44.4^\circ)}{\sin(43.8^\circ)}\right]^2 = 1.02. \tag{7.8}$$

Problem 7.4 *About* 10^6 *electron-ion pairs are produced by a charged particle traversing a counter. If the typical ionization potential of the medium is* $\bar{I} = 30\,\text{eV}$, *in principle, how well can you measure the deposited energy using a Geiger counter, an ionization* *counter with a gain of unity, and a proportional counter with a gain of* 10^6 *that has gain variations of* 5%?

The statistical uncertainty (in %) is related to the number of counts detected by the counter

$$\sigma_{\text{statistical}} = \frac{1}{\sqrt{n}}. \tag{7.9}$$

With 30 eV/ion-pair and 10^6 produced ion pairs, 30 MeV is deposited in the counter. A Geiger Counter will provide only a single count for each particle detected and therefore has no resolution for the amount of ionization. An Ionization Counter detecting single ionization electrons could in principle reach a resolution of

$$\sigma = \frac{1}{\sqrt{n}} = \frac{1}{\sqrt{1 \times 10^6}} = 0.1\% \text{ or } 30\,\text{keV}.$$

However, electronic noise associated with the environment and the readout electronics corresponds to, at best, a few thousand electrons, and thus dominates the energy resolution. A Proportional Counter producing 10^{12} ion pairs has a negligible statistical error, however, gain variations of $\approx 5\%$ generate an observed uncertainty of $\approx 5\%$, or 1.5 MeV.

Problem 7.5 *If you wish to measure the momentum of a* 10 GeV/c *singly-charged particle to* 1% *accuracy, in a* 2 T *field, using a* 1 m *long magnet, how well do you have to know the exit angle (see Fig. 7.5)? If you use MWPCs that have* 2 mm *inter-wire anode spacings to measure that angle, about how far do you have to separate two planes to achieve your goal? Now suppose that you use, instead, silicon microstrip detectors of* 25 μm *spacing. What separation distance between two such planes could achieve the same goal?*

A magnetic field $B = 2\,\text{T}$ with an effective length $L = 1\,\text{m}$, alters the trajectory of a singly-charged particle with momentum $p = 10\,\text{GeV}/c$

as follows. The magnet imparts a transverse momentum

$$p_T = BL \tag{7.10}$$

relative to the incident direction, and the particle will therefore leave the field region at an angle given by

$$\sin\theta = \frac{p_T}{p} = \frac{BL}{p} = \frac{(2\,\mathrm{T})(1\,\mathrm{m})}{19} = 0.2. \tag{7.11}$$

A hit on any MWPC wire for a wire pitch $d = 2\,\mathrm{mm}$ stipulates the location of the particle with uniform probability of being within a width $\pm d/2$ of the hit location.

The resolution of this measurement is taken as the standard deviation of the uniform probability distribution (better than using just $d/2$):

$$\sigma = d/\sqrt{12} = \frac{2\,\mathrm{mm}}{\sqrt{12}} = 0.58\,\mathrm{mm}. \tag{7.12}$$

Two points separated by a distance D along the flight path, each measured with a resolution σ, define the angular resolution

$$\sigma_\theta = \sqrt{2}\frac{\sigma}{D}. \tag{7.13}$$

To measure a momentum to 1% accuracy, the bend angle θ must be measured to the same accuracy

$$d\theta = (0.01)\theta = 0.002\ \mathrm{rad} = 2\,\mathrm{mrad}. \tag{7.14}$$

Equating the last two equations yields:

$$D = \sqrt{2}\sigma_\theta/d\theta = (1.4)(0.58\,\mathrm{mm})/2\times 10^{-3} = 40\,\mathrm{cm}. \tag{7.15}$$

For silicon detectors with strip pitch of $d' = 0.025\,\mathrm{mm}$, the equivalent spacing of planes is much smaller:

$$D' = (d'/d)D = (0.025/2)(40\,\mathrm{cm}) = 0.48\,\mathrm{cm}. \tag{7.16}$$

Problem 7.6 *Sketch the pulse height spectrum that you would expect in the decay of* ^{60}Co *in Eq.* (7.8) *when the two de-excitation photons are emitted simultaneously, namely within the time resolution of the detector.*

The energy deposited in each event is the sum of the photon energies $= 2.5\,\mathrm{MeV}$, with a fractional uncertainty characteristic of the detector, namely 10% or $\sigma = 0.25\,\mathrm{MeV}$.

8. Accelerators

Problem 8.1 *Protons are accelerated in a cyclotron by an electric field with oscillating frequency of* 8 MHz. *If the diameter of the magnet is* 1 m, *calculate the value of magnetic field and the maximum energy that the protons can reach.*

Protons are accelerated in a cyclotron of radius 0.5 m, and an 8 MHz acceleration frequency ($\omega = 2\pi f = 5 \times 10^7$ rad/s). Protons will circulate at an 8 MHz frequency for a magnetic field of:

$$B = \frac{m\omega}{Q} = \frac{(1.67 \times 10^{-27}\ \text{kg})(5.0 \times 19^7\ \text{rad/s})}{1.6 \times 10^{-19}\text{C}} = 0.52\ \text{T} \tag{8.1}$$

or, using eV units for the mass

$$\begin{aligned} B &= \frac{mc^2\omega}{Qc^2} = \left(\frac{mc^2}{Q}\right)\frac{\omega}{c^2} \\ &= (940 \times 10^6\,\text{eV})\frac{(5.0 \times 10^7\,\text{rad/s})}{9 \times 10^{16}\,\text{m}^2/\text{s}^2} = 0.52\,\text{T}. \end{aligned} \tag{8.2}$$

The maximum kinetic energy

$$\begin{aligned} T &= \frac{1}{2}mv^2 = \frac{1}{2}\left(mc^2\right)\left(\frac{\omega R}{c}\right)^2 \\ &= 0.5(940\,\text{MeV})\left[\frac{2.5 \times 10^7}{3 \times 10^8}\right]^2 = 3.3\,\text{MeV}. \end{aligned} \tag{8.3}$$

Problem 8.2 *To achieve an energy of* 20 TeV, *each of the SSC main rings was to contain about* 4000 *dipole magnets, each* 16*-meters long, with a field of* 7 T. *This means that over half of the* ≈ 60 *mile SSC tunnel was to be taken up by dipoles. If you were to build a single synchrotron for use in fixed-target collisions of equivalent energy in* *the center-of-mass* ($\sqrt{s} = 40\,\text{TeV}$), *and used a similar magnet design, how long would your tunnel have to be?*

The SSC was designed to have center mass energy of 40 TeV. A fixed-target machine with a beam energy E' has a center-of-mass energy

$$\sqrt{s} = \sqrt{2mE'} \tag{8.4}$$

to have the same center-of-mass energy as the SSC, the energy of a fixed target beam must be:

$$E' = \frac{(\sqrt{2})^2}{2m_p c^2} = \frac{1600 \times 10^6\,\text{GeV}^2}{2\,\text{GeV}} = 9 \times 10^5\,\text{TeV}. \tag{8.5}$$

The circumference of the circular tunnel will scale with the energy

$$C' = C\frac{E'}{E} = (60\,\text{miles})\frac{8 \times 10^5\,\text{TeV}}{20\,\text{TeV}} = 2.4 \times 10^6\,\text{miles}. \tag{8.6}$$

Problem 8.3 *If the capacitance of a Van de Graaff accelerator terminal is* $250\,\mu\mu F$ *(pF), and if it operates at a voltage of* $4\,\text{MV}$, *what is the total charge on the terminal? If the charging belt can carry a current of* $0.2\,\text{mA}$, *how long does it take to charge up the accelerator to* $4\,\text{MV}$?

The total charge on a capacitance C at a voltage V is:

$$Q = CV = (2.5 \times 10^{-10}\,\text{F})(4 \times 10^6\,\text{V}) = 10^{-3}\,\text{C}. \tag{8.7}$$

A constant current I will charge the capacitor in a time

$$t = \frac{Q}{I} = \frac{10^{-3}\,\text{C}}{0.2 \times 10^{-3}\,\text{C/s}} = 5\,\text{s}. \tag{8.8}$$

Problem 8.4 *Starting with cgs units, show that Relation* (8.9′) *follows from Relation* (8.9).

Starting from the relationship in SI unit for the momentum p of a particle with charge Q circulating at a radius R in a magnetic field B, the relationship in mixed-units (Relation (8.9′)) can be obtained as follows:

$$p = QBR \cdot [1(\text{kg} \cdot \text{m/s})/(\text{CTm})]. \tag{8.9}$$

Inserting a ratio for the speed of light and electron charge:

$$p = QBR \cdot \left[\left(\frac{3 \times 10^8\,\text{m/s}}{c} \right) (\text{kg} \cdot \text{m/s}) \left(\frac{1.6 \times 10^{-19}\text{C}}{1.6 \times 10^{-19}\text{C}} \right) \Big/ (\text{CTm}) \right].$$

Note that $1.6 \times 10^{-19}\,\text{kg} \cdot \text{m}^2/\text{s}^2 = 1.6 \times 10^{-19}\,\text{J} = 1\,\text{eV}$, and introducing $Q/e = q$

$$p = qBR[(3 \times 10^8)(\text{eV}/c)/(\text{Tm})]. \tag{8.10}$$

Finally, changing to GeV energy units, where $3 \times 10^8\,\text{eV} = 0.3\,\text{GeV}$, we get:

$$p = 0.3\,qBR[(\text{GeV}/c)/(\text{Tm})]. \tag{8.11}$$

Problem 8.5 *Suggest a mechanism whereby an accelerated beam could be extracted from a circular accelerator, and directed onto an external target.*

Except for a pulsed magnet, all other techniques require the orbit to be "bumped" into a region of special magnetic or electric field. In a magnetic field, an electrostatic septum can be used, where one thin plate (or plane of wires) of a capacitor separates a region of E-field from an E-field-free region. In a straight section, either an electrostatic septum, or a magnetic septum providing a (thin) current sheet that separates a region of B-field from a B-field-free region, can be used.

Problem 8.6 *Using Eq.* (8.12), *there is ostensibly sufficient energy in the center of mass in the collision of a* $1\,\text{TeV}\,\pi^0$ *with a lead nucleus at rest to produce a Higgs boson* (H^0) *of* $M_H \approx 120\,\text{GeV}/c^2$. *In principle, this can be done in a coherent collision, where the Pb nucleus remains intact. Does this make sense in light of Footnote 1? Assuming a nuclear form factor for Pb of* $\approx e^{-400q^2}$ (*with* q *in* GeV *units*), *and considering the silly reaction* $\pi^0 + Pb \to H^0 + Pb$, *what would be the approximate reduction in the probability for producing the Higgs at* $0°$ *as a result of the form factor?* [*Hints: To calculate the minimum value allowed for the quantity*

$q^2 = (\vec{p}_\pi c - \vec{p}_H c)^2 - (E_\pi - E_H)^2$, assume that the Higgs boson is relativistic, but approximate the terms to order $(M_H c^2/E_H)^4$ in β. You should get that $q^2 \approx q^2_{\min} \approx (M_H^2 c^4/2E_\pi)^2$, when you ignore the small mass of the pion and set $E_H = E_\pi$.]

Coherent Higgs (m_H : 120 GeV/c^2) production via the interaction of a 1 TeV π^0 in the Coulomb field of a Pb nucleus at rest is pictured in the accompanying Fig. 1.

The Coulomb field of a nucleus extends significantly beyond the size of the nucleus, so that, in principle, this reaction can be coherent, i.e. involve the charge of the entire nucleus. It is a stretch of the imagination to believe that a neutral π^0 that decays in a proper time of $\sim 10^{-16}$ seconds will have any time to interact in this way, but let us continue this amusing exercise:

$$p_H c = \sqrt{E_H^2 - m_H^2 c^4} = E_H \left(1 - \frac{m_H^2 c^4}{E_H^2}\right)^{\frac{1}{2}}$$

$$= E_H \left(1 - \frac{m_H^2 c^4}{2E_H^2} - \frac{m_H^4 c^4}{8E_H^4} + K\right), \qquad (8.12)$$

where K refers to higher-order terms in the expansion.

Assuming that the Higgs has the same direction as the incident π^0, the square in the four-momentum transfer is:

$$q^2 = (p_\pi c - p_H c)^2 - (E_\pi - E_H)^2$$
$$= p_\pi^2 c^2 - 2p_\pi p_H c^2 + p_H^2 c^2 - E_\pi^2 + 2E_\pi E_H - E_H^2. \qquad (8.13)$$

Making the approximations $E_\pi \approx p_\pi c$ and $\cos\theta_{\pi H} \approx 1$, and using the expansion in Eq. (8.12) above,

$$q^2 = m_\pi^2 c^4 - m_H^2 c^4 + 2E_\pi(E_H - p_H c)$$
$$\approx -m_H^2 c^4 + 2E_\pi \left[E_H - E_H\left(1 - \frac{m_H^2 c^4}{2E_H^2} - \frac{M_H^4 c^4}{8E_H^4} + K\right)\right]. \qquad (8.14)$$

Taking $E_\pi \approx E_H$, only the third term in the expansion survives, yielding

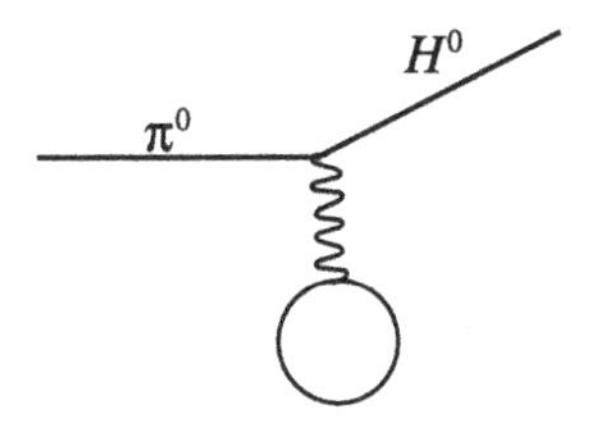

Fig. 8.1. Totally unrealistic mechanism for producing a Higgs boson in coherent scattering of neutral pions from Pb nuclei.

$$q^2 = -m_H^2 c^4 + 2E_H^2 \left[+ \frac{m_H^2 c^4}{2E_H^2} + \frac{M_H^4 c^8}{8E_H^4} \right] \approx \frac{m_H^4 c^8}{4E_H^2}. \tag{8.15}$$

Calculating the momentum transfer for the given mass and energy

$$q^2 = \left(\frac{m_H^2 c^4}{2E_H} \right)^2 = \left(\frac{14400}{2000} \right)^2 = 50\,\mathrm{GeV}^2. \tag{8.16}$$

The form factor would therefore yield an infinitesimal probability for the process

$$e^{-400q^2} = e^{-20000} \approx 0. \tag{8.17}$$

9. Properties and Interactions of Elementary Particles

Problem 9.1 *What quantum numbers, if any, are violated in the following reactions? Are the interactions strong, weak, electromagnetic, or none of the above? (See the CRC Handbook for particle properties.)*

(*a*) $\Omega^- \to \Xi^0 + \pi^-$,
(*b*) $\Sigma^+ \to \pi^+ + \pi^0$,
(*c*) $n \to p + \pi^-$,
(*d*) $\pi^0 \to \mu^+ + e^- + \bar{\nu}_e$,
(*e*) $K^0 \to K^+ + e^- + \bar{\nu}_e$,
(*f*) $\Lambda^0 \to p + e^-$.

(a) Let us consider the reaction

$$\Omega^- \to \Xi^0 + \pi^-. \tag{9.1}$$

From Table 9.4 of the text, we note that

$$I_3(\Omega^-) = 0, \quad S(\Omega^-) = -3, \quad I_3(\pi^-) = -1, \quad S(\pi^-) = 0. \tag{9.2}$$

Similarly, from the reaction in Eq. (9.22) of the text, we know that

$$I_3(\Xi^0) = \frac{1}{2}, \quad S(\Xi^0) = -2. \tag{9.3}$$

Using these quantum numbers in Eq. (9.1), we find that the reaction violates both isospin and strangeness quantum numbers:

$$|\Delta I_3| = \frac{1}{2}, \quad |\Delta S| = 1. \tag{9.4}$$

This is a weak hadronic decay.

(b) For the reaction

$$\Sigma^+ \to \pi^+ + \pi^0, \tag{9.5}$$

we note from Table 9.4 of the text that

$$\begin{aligned} I_3(\Sigma^+) &= 1, \quad I_3(\pi^+) = 1, \quad I_3(\pi^0) = 0, \\ B(\Sigma^+) &= 1, \quad B(\pi^+) = B(\pi^0) = 0, \\ S(\Sigma^+) &= -1, \quad S(\pi^+) = S(\pi^0) = 0. \end{aligned} \tag{9.6}$$

Using these, we find that Reaction (9.5) violates baryon number as well as strangeness number:

$$|\Delta B| = 1, \quad |\Delta S| = 1. \tag{9.7}$$

Baryon number violating processes have not yet been observed, and this reaction cannot be classified as a strong, weak or electromagnetic process.

(c) The reaction

$$n \to p + \pi^-, \tag{9.8}$$

can be checked from Table 9.4 to satisfy conservation of all the quantum numbers. However, from Table 9.3 of the text we see that while the mass difference between the neutron and proton is about 1.3 MeV/c^2, the mass of π^- is about 140 MeV/c^2. Thus, such a reaction will violate energy–momentum conservation and, consequently, is not an allowed reaction since energy–momentum conservation is absolute. Nevertheless, this reaction can take place virtually (as for the case of $\gamma \to e^+e^-$) in the strong or Coulomb nuclear field (see, e.g., Fig. 6.4 in the text).

(d) The decay of the π^0 meson through

$$\pi^0 \to \mu^+ + e^- + \bar{\nu}_e, \tag{9.9}$$

violates lepton number conservation. More precisely, from Table 9.2 of the text we see that the muon lepton number is violated in this process. This kind of decay has not as yet been observed, and does not fall into strong, weak or electromagnetic processes.

(e) The decay

$$K^0 \to K^+ + e^- + \bar{\nu}_e, \tag{9.10}$$

is kinematically allowed, as we see from Table 9.3 of the text. From the quantum numbers in Table 9.4 of the text we have

$$I_3(K^0) = -\frac{1}{2}, \quad I_3(K^+) = \frac{1}{2}, \quad S(K^0) = 1, \quad S(K^+) = 1. \tag{9.11}$$

The reaction therefore violates isospin, but not strangeness in the hadronic sector (which is the meaningful sector for assigning strong quantum numbers, as discussed in Sec. 9.8.1 of the text):

$$|\Delta I_3| = 1, \quad |\Delta S| = 0. \tag{9.12}$$

This can be classified as a weak semi-leptonic decay.

(f) From Tables 9.2 and 9.4 of the text, we note that the decay process

$$\Lambda^0 \to p + e^-, \tag{9.13}$$

violates lepton number, isospin and strangeness number

$$|\Delta L| = 1, \quad |\Delta I_3| = \frac{1}{2}, \quad |\Delta S| = 1. \tag{9.14}$$

This reaction has never been observed, and does not fall into the strong, weak or the electromagnetic category of reactions.

Problem 9.2 *What quantum numbers, if any, are violated in the following processes? Would the reaction be strong, electromagnetic, weak, or unusually suppressed? Explain. (See CRC Handbook for particle properties.)*

(a) $\Lambda^0 \to p + e^- + \bar{\nu}_e$,
(b) $K^- + p \to K^+ + \Xi^-$,
(c) $K^+ + p \to K^+ + \Sigma^+ + \bar{K}^0$,
(d) $p + p \to K^+ + K^+ + n + n$,
(e) $\Sigma^+(1385) \to \Lambda^0 + \pi^+$,
(f) $\bar{p} + n \to \pi^- + \pi^0$.

(a) From Table 9.4 we have that

$$I_3(\Lambda^0) = 0, \quad I_3(p) = \frac{1}{2}, \quad S(\lambda^0) = -1, \quad S(p) = 0. \tag{9.15}$$

Therefore, we see that the hadronic sector of the decay in

$$\Lambda^0 \to p + e^- + \bar{\nu}_e, \tag{9.16}$$

violates both isospin and strangeness quantum numbers

$$|\Delta I_3| = \frac{1}{2}, \quad |\Delta S| = 1. \tag{9.17}$$

This represents a weak semi-leptonic decay process.
(b) The process

$$K^- + p \to K^+ + \Xi^- \tag{9.18}$$

is seen from Table 9.4 of the text to satisfy conservation of all quantum numbers. This is indeed a strong process.
(c) We note from Table 9.4 of the text that

$$\begin{aligned} I_3(K^+) &= \frac{1}{2}, \quad I_3(p) = I_3(\bar{K}^0), \qquad I_3(\Sigma^+) = 1, \\ S(K^+) &= 1, \quad S(\bar{K}^0) = -1 = S(\Sigma^+), \quad S(p) = 0. \end{aligned} \tag{9.19}$$

Therefore, the reaction

$$K^+ + p \to K^+ + \Sigma^+ + \bar{K}^0 \tag{9.20}$$

violates both isospin and strangeness quantum numbers

$$|\Delta I_3| = 1, \quad |\Delta S| = 2. \tag{9.21}$$

Because strangeness changes by two units, it is a highly suppressed unknown hadronic process.
(d) As in the earlier reaction, the process

$$p + p \to K^+ + K^+ + n + n \tag{9.22}$$

is also a highly suppressed unknown hadronic reaction. We note that

$$\begin{aligned} I_3(p) &= \frac{1}{2}, \quad I_3(K^+) = \frac{1}{2}, \quad I_3(n) = -\frac{1}{2}, \\ S(p) &= 0, \quad S(K^+) = 1, \quad S(n) = 0, \end{aligned} \tag{9.23}$$

so that the process (9.22) satisfies

$$|\Delta I_3| = 1, \quad |\Delta S| = 2. \tag{9.24}$$

(e) From Table 9.4 of the text, we see that the decay

$$\Sigma^+(1385) \rightarrow \Lambda^0 + \pi^+ \tag{9.25}$$

conserves all quantum numbers. It is a strong process. (Note that this channel is not kinematically allowed in $\Sigma^+(1189.4)$ decay.)

(f) The process

$$\bar{p} + n \rightarrow \pi^- + \pi^0, \tag{9.26}$$

involves only nonstrange hadrons and from Table 9.4 appears to satisfy conservation of all quantum numbers. This is a strong process. (The inverse reaction would only be allowed virtually.)

Problem 9.3 *A π^0 meson with momentum* 135 GeV/c *decays into two photons. If the mean life of a π^0 is* 8.5×10^{-17} sec, *calculate to* 10% *accuracy how far the high-energy π^0 will travel prior to decay. What will be the approximate minimum value of the opening angle of its two decay photons in the laboratory?*

The π^0 meson has momentum

$$p = 135\,\text{GeV}/c = 135 \times 10^3\,\text{MeV}/c, \tag{9.27}$$

and from Table 9.3 of the text, it has a rest mass

$$m = 135\,\text{MeV}/c^2, \tag{9.28}$$

so that we can write

$$p = 10^3 mc. \tag{9.29}$$

The π^0 meson is therefore ultrarelativistic, and we obtain

$$E = \sqrt{p^2c^2 + m^2c^4} = \sqrt{10^6 + 1}\; mc^2 \approx 10^3 mc^2 = pc. \tag{9.30}$$

Comparing with Eq. (A.7) of the text we determine

$$\begin{aligned} E &= \gamma mc^2, && \Rightarrow \gamma \approx 10^3, \\ p &= 10^3 mc = \gamma\beta mc, && \Rightarrow \beta \approx 1. \end{aligned} \tag{9.31}$$

In its rest frame, the π^0 meson has a mean life

$$\tau^{(\text{rest})} = 8.5 \times 10^{-17}\,\text{sec}. \tag{9.32}$$

When the π^0 meson is moving, there is time dilation leading to the mean life time (see Eq. (A.13) of the text)

$$\tau = \gamma\tau^{(\text{rest})} \approx 10^3 \times 8.5 \times 10^{-17}\,\text{sec} = 8.5 \times 10^{-14}\,\text{sec}. \tag{9.33}$$

During this time interval, the π^0 meson will traverse a distance

$$\begin{aligned} d &= \beta c\tau \approx 1 \times 3 \times 10^{10}\,\text{cm/sec} \times 8.5 \times 10^{-14}\,\text{sec} \\ &= 2.55 \times 10^{-3} \approx 0.025\,\text{mm} = 25\,\mu\text{m}, \end{aligned} \tag{9.34}$$

which is hard but not impossible to measure.

In the decay of the π^0 meson, for the two photons emitted at some opening angle θ, energy–momentum conservation leads to

$$\begin{aligned} \vec{p} &= \vec{p}_1 + \vec{p}_2, \\ E &= E_1 + E_2 = p_1c + p_2c \\ \text{or} \quad \frac{E}{c} &= p_1 + p_2, \end{aligned} \tag{9.35}$$

where we have used the fact that for massless photons the Einstein relationship takes the form

$$E_i = p_i c, \quad i = 1, 2. \tag{9.36}$$

Squaring the two relationships in Eq. (9.36) and subtracting one from the other, we obtain

$$\begin{aligned} & E^2 - p^2c^2 = (E_1 + E_2)^2 - (\vec{p}_1 + \vec{p}_2)^2c^2 \\ \text{or}\ & (E^2 - p^2c^2) = (2p_1p_2 - 2\vec{p}_1 \cdot \vec{p}_2)c^2 \\ \text{or}\ & m^2c^4 = 2E_1E_2(1 - \cos\theta) = 2E_1(E - E_1)(1 - \cos\theta) \\ \text{or}\ & \cos\theta = 1 - \frac{m^2c^4}{2E_1(E - E_1)}. \end{aligned} \tag{9.37}$$

Clearly, the opening angle is a function of E_1 (everything else is fixed), and the minimum opening angle can be determined by taking the derivative of the above expression with respect to E_1 and setting it to zero, which leads to

$$\frac{m^2c^4}{2}\left(\frac{1}{E_1^2(E-E_1)} - \frac{1}{E_1(E-E_1)^2}\right) = 0$$

$$\text{or} \quad \frac{m^2c^4}{2E_1^2(E-E_1)^2}(E-E_1-E_1) = 0 \tag{9.38}$$

$$\text{or} \quad E_1 = \frac{E}{2} = E_2.$$

In turn, this leads from Eq. (9.38) to

$$\begin{aligned}\cos\theta_{\text{min}} &= 1 - \frac{m^2c^4}{2\left(\frac{E}{2}\right)^2} = 1 - \frac{2m^2c^4}{E^2} \approx 1 - \frac{2m^2c^4}{p^2c^2} \\ &= 1 - \frac{2}{(10^3)^2} = 1 - 2\times 10^{-6}.\end{aligned} \tag{9.39}$$

It is clear that when θ_{min} is extremely small, as it is here, we can Taylor expand $\cos\theta_{\text{min}}$ to obtain

$$\begin{aligned} 1 - \frac{(\theta_{\text{min}})^2}{2} &\approx 1 - \frac{2m^2c^4}{E^2} = 1 - 2\times 10^{-6} \\ \text{or} \quad \theta_{\text{min}} &\approx \frac{2mc^2}{E} \approx 2\times 10^{-3}\,\text{rad} = 2\,\text{mrad}.\end{aligned} \tag{9.40}$$

Problem 9.4 *We will see in Chap. 13 that hadrons are composed of constituents known as quarks, and that mesons can be represented as quark–antiquark systems and baryons as three-quark systems. All quarks have baryon number $\frac{1}{3}$, and their other quantum numbers are listed in Table 9.5. Antiquarks have all their quantum numbers of opposite sign to the quarks. The isotopic spin of quarks can be inferred from the generalized Gell-Mann–Nishijima relation of Eq. (9.26). Free quarks are not observed in nature. The top quark is as free as a quark can get, but it decays so rapidly that it does not have sufficient time to form hadrons, reflecting the fact that its weak interactions are stronger than its strong interactions.*

The quark system uds can exist in more than one isospin state. What is the value of I_3 for this combination of quarks? What are the possible values of total I-spin for uds states? Can you identify them with any known particles? (See, e.g., CRC Handbook.)

Table 9.5. Properties of the quarks

Quark	Symbol	Rest mass (GeV/c^2)	Electric charge (e)	"Flavor" Quantum Numbers			
				Strange	Charm	Bottom	Top
Up	u	$\lesssim 3\times10^{-3}$	$\frac{2}{3}$	0	0	0	0
Down	d	$\approx 7\times10^{-3}$	$-\frac{1}{3}$	0	0	0	0
Strange	s	≈ 0.12	$-\frac{1}{3}$	−1	0	0	0
Charm	c	≈ 1.2	$\frac{2}{3}$	0	1	0	0
Bottom	b	≈ 4.2	$-\frac{1}{3}$	0	0	−1	0
Top	t	175 ± 5	$\frac{2}{3}$	0	0	−0	1

Since quarks carry baryon number

$$N = \frac{1}{3}, \tag{9.41}$$

from the Gell-Mann–Nishijima relationship in Eq. (9.26) of the text

$$Q = I_3 + \frac{N+S}{2}, \tag{9.42}$$

as well as Table 9.5 of the text, we see that

$$\begin{aligned}
I_3(u) &= Q(u) - \frac{N(u)+S(u)}{2} = \frac{2}{3} - \frac{\frac{1}{3}+0}{2} = \frac{2}{3} - \frac{1}{6} = \frac{1}{2},\\
I_3(d) &= Q(d) - \frac{N(d)+S(d)}{2} = -\frac{1}{3} - \frac{\frac{1}{3}+0}{2} = -\frac{1}{3} - \frac{1}{6} = -\frac{1}{2},\\
I_3(s) &= Q(s) - \frac{N(s)+S(s)}{2} = -\frac{1}{3} - \frac{\frac{1}{3}-1}{2} = -\frac{1}{3} + \frac{1}{3} = 0.
\end{aligned} \tag{9.43}$$

Thus, the u, d quarks have isospin $I = \frac{1}{2}$, while the strange quark is an isospin singlet $I = 0$. The resulting isospin for a composite state such as (uds) can be determined from the law of addition of angular momentum (isospin algebra is like angular-momentum algebra). Thus, two $I = \frac{1}{2}$ objects can form an $I = 0$ or an $I = 1$ object, which can be combined with another $I = 0$ to yield:

$$I(uds) = 0 \text{ and } 1. \tag{9.44}$$

The (uds) state is charge neutral, has $N = 1$ and $S = -1$. Comparing with the properties of the particles in Table 9.4 of the text, we conclude that the different isospin states can be identified with the particles

$$\begin{aligned} I = 0 &: \Lambda^0 \text{ and the family of } \Lambda^0 \text{ particles,} \\ I = 1 &: \Sigma^0 \text{ and the family of } \Sigma^0 \text{ particles.} \end{aligned} \tag{9.45}$$

Problem 9.5 *What is the baryon number, hypercharge, and isotopic spin of the following quark systems:* (a) $u\bar{s}$, (b) $c\bar{d}$, (c) $\bar{u}\bar{u}\bar{d}$, (d) ddc, (e) ubc, (f) $s\bar{s}$. *Using the CRC Handbook, can you identify these states with any known particles?*

In the last problem we determined

$$I_3(u) = \frac{1}{2}, \quad I_3(d) = -\frac{1}{2}, \quad I_3(s) = 0, \quad I(u) = I(d) = \frac{1}{2}, \quad I(s) = 0. \tag{9.46}$$

For particles with flavor quantum numbers beyond strangeness, the Gell-Mann–Nishijima relationship holds with a generalization of the definition of hypercharge to include the new flavor quantum numbers, namely,

$$Y = N + S + C + B + T, \quad Q = N + \frac{Y}{2}, \tag{9.47}$$

where C, B, and T represent the Charm, Bottom and Top quantum numbers. Using this generalized definition, we can now calculate

$$\begin{aligned} I_3(c) &= Q(c) - \frac{Y(c)}{2} = \frac{2}{3} - \frac{\frac{1}{3} + 0 + 1 + 0 + 0}{2} = \frac{2}{3} - \frac{2}{3} = 0, \\ I_3(b) &= Q(b) - \frac{Y(b)}{2} = -\frac{1}{3} - \frac{\frac{1}{3} + 0 + 0 - 1 + 0}{2} = -\frac{1}{3} + \frac{1}{3} = 0, \\ I_3(t) &= Q(t) - \frac{Y(t)}{2} = \frac{2}{3} - \frac{\frac{1}{3} + 0 + 0 + 0 = 1}{2} = \frac{2}{3} - \frac{2}{3} = 0. \end{aligned} \tag{9.48}$$

Thus, we see that u and d belong to the $I = \frac{1}{2}$ representation of isospin, while the quarks with exotic flavors are isospin singlets with

$I = 0$. The antiquarks belong to the same isospin representations, but have opposite values for all other quantum numbers. With these facts in mind, let us look at some composite states.

(a) The state

$$u\bar{s} \tag{9.49}$$

is a charged state with

$$N = \frac{1}{3} - \frac{1}{3} = 0, \quad Q = \frac{2}{3} + \frac{1}{3} = 1, \quad S = 0 + 1 = 1, \tag{9.50}$$

with vanishing quantum numbers for $C = B = T = 0$. This is, therefore, a charged, strange-meson state and belongs to the isospin representation of an $I = \frac{1}{2}$ interacting with an $I = 0$, or

$$I(u\bar{s}) = \frac{1}{2}, \tag{9.51}$$

with

$$I_3(u\bar{s}) = I_3(u) + I_3(\bar{s}) = \frac{1}{2} + 0 = \frac{1}{2}. \tag{9.52}$$

From Table 9.4 of the text, we see that this state corresponds to the particle K^+ or its excited states.

(b) The composite state $(c\bar{d})$ has the quantum numbers

$$N = \frac{1}{3} - \frac{1}{3} = 0, \quad Q = \frac{2}{3} + \frac{1}{3} = 1, \quad C = 1 + 0 = 1, \tag{9.53}$$

with vanishing quantum numbers $S = B = T = 0$. This is therefore also a charged meson with charm quantum number. It belongs to the isospin representation

$$I(c\bar{d}) = \frac{1}{2}, \tag{9.54}$$

with

$$I_3(c\bar{d}) = I_3(c) + I_3(\bar{d}) = 0 + \frac{1}{2} = \frac{1}{2}. \tag{9.55}$$

Comparing with the particles in the *CRC Handbook* or the *Particle Data Group* (*PDG*) book, we conclude that this state corresponds to the D^+ meson or its excited states.

(c) The state $(\bar{u}\bar{u}\bar{d})$ is the antiparticle state of (uud) and has the quantum numbers

$$N = -\frac{1}{3} - \frac{1}{3} - \frac{1}{3} = -1, \quad Q = -\frac{2}{3} - \frac{2}{3} + \frac{1}{3} = -1, \tag{9.56}$$

with $S = C = B = T = 0$. Thus, it is a charged, nonstrange antibaryon. Because $\bar{u}$ and $\bar{d}$ have $I = \frac{1}{2}$, the $(\bar{u}\ \bar{d})$ system can be either $I = 0$ or $I = 1$. Thus, the $(\bar{u}\bar{u}\bar{d})$ state can be either:

$$I(\bar{u}\bar{u}\bar{d}) = \frac{1}{2} \text{ or } \frac{3}{2} \tag{9.57}$$

with

$$I_3(\bar{u}\bar{u}\bar{d}) = I_3(\bar{u}) + I_3(\bar{u}) + I_3(\bar{d}) = -\frac{1}{2} - \frac{1}{2} + \frac{1}{2} = -\frac{1}{2}. \tag{9.58}$$

The $I = \frac{1}{2}$ state corresponds to $\bar{p}$ (antiproton) or its excited states, while the $I = \frac{3}{2}$ state can be identified with $\bar{\Delta}^-$ or its excited states.

(d) The composite state (ddc) has the quantum numbers

$$N = \frac{1}{3} + \frac{1}{3} + \frac{1}{3} = 1, \quad Q = -\frac{1}{3} - \frac{1}{3} + \frac{2}{3} = 0, \quad C = 0 + 0 + 1 = 1, \tag{9.59}$$

with $S = B = T = 0$. Thus, this is a charge-neutral charmed baryon state. Arguing as before, it belongs to the isospin representations:

$$I(ddc) = 0, 1 \tag{9.60}$$

with

$$I_3(ddc) = I_3(d) + I_3(d) + I_3(c) = -\frac{1}{2} - \frac{1}{2} + 0 = -1. \tag{9.61}$$

From the *CRC Handbook* or the *PDG* book, we conclude that the $I = 0$ representation corresponds to the Λ_c^0 particle and its excited states, while the $I = 1$ representation corresponds to the Σ_c^0 particle and its excited states.

(e) The composite state (ubc) has the quantum numbers

$$N = \frac{1}{3} + \frac{1}{3} + \frac{1}{3} = 1, \quad Q = \frac{2}{3} - \frac{1}{3} + \frac{2}{3} = 1,$$
$$C = 0 + 0 + 1 = 1, \qquad B = 0 - 1 + 0 = -1, \tag{9.62}$$

with $S = 0$. This is a nonstrange charged baryon, with charm and bottom quantum numbers. It belongs to the isospin representation

$$I(ubc) = \frac{1}{2}, \tag{9.63}$$

with

$$I_3(ubc) = I_3(u) + I_3(b) + I_3(c) = \frac{1}{2} + 0 + 0 = \frac{1}{2}. \tag{9.64}$$

These types of states have not yet been confirmed.

(f) The state $(s\bar{s})$ has the quantum numbers

$$N = \frac{1}{3} - \frac{1}{3} = 0, \quad Q = -\frac{1}{3} + \frac{1}{3} = 0, \tag{9.65}$$

with $S = C = B = T = 0$. Namely, it is a nonstrange, charge-neutral meson state without flavor. Because s quarks have $I = 0$, it belongs to the isospin representation

$$I(s\bar{s}) = 0, \tag{9.66}$$

with

$$I_3(s\bar{s}) = I_3(s) + I_3(\bar{s}) = 0 + 0 = 0. \tag{9.67}$$

From Table 9.4 of the text, we see that this could correspond to the $J = 0$ η^0 particle and its excited states. In fact, the $J = 1$ meson ϕ is thought to be an almost pure $(s\bar{s})$ state (see the *CRC Handbook*).

Problem 9.6 *Consider the following decays:*

(a) $N^+(1535) \rightarrow p + \eta^0$,
(b) $\Sigma^+(1189) \rightarrow p + \pi^0$,
(c) $\rho^0(770) \rightarrow \pi^0 + \gamma$.

(a) From the *CRC Handbook* or the *PDG* booklet, we note that $N^+(1535)$ is a nonstrange baryon with isospin $I = \frac{1}{2}, I_3 = \frac{1}{2}$. Thus, we see that the reaction

$$N^+(1535) \to p + \eta^0 \tag{9.68}$$

conserves all quantum numbers and is therefore a strong process. The spin of the N^+ is $J = \frac{1}{2}$, while $J_p = \frac{1}{2}$, and $J_\eta = 0$. Consequently, we can have $\ell = 0$ and $\ell = 1$ as relative orbital-momentum quantum numbers in the final state. But only $\ell = 1$ is allowed for strong interactions (see Chap. 11).

(b) The process

$$\Sigma^+(1189) \to p + \pi^0 \tag{9.69}$$

can be checked from Table 9.4 of the text to violate both isospin and strangeness quantum numbers:

$$|\Delta I_3| = \left|1 - \frac{1}{2} - 0\right| = \frac{1}{2}, \qquad |\Delta S| = |-1 - 0 - 0| = 1. \tag{9.70}$$

It is therefore a weak hadronic process. As in part (a), both $\ell = 0$ and $\ell = 1$ can occur in the final state, and, in fact, "parity" violation in weak interactions allows both (see Chap. 11).

(c) From the *CRC Handbook* or the *PDG* book, we note that the $\rho^0(770)$ meson has $I = 1, I_3 = 0$. Therefore, the process

$$\rho^0(770) \to \pi^0 + \gamma \tag{9.71}$$

conserves all quantum numbers. It may seem then that this would correspond to a strong interaction. However, the photon participates only in electromagnetic processes (and not in strong reactions). We therefore conclude that this is an electromagnetic transition. The ρ^0 decays mainly into $\pi^+\pi^-$ pairs, and only $\approx 10^{-3}$ of the time to $\pi^0 + \gamma$. The ρ has $J = 1$, while $J_{\pi^0} = 0$. The photon must therefore carry away $J = 1$. Electromagnetic interactions conserve parity, and the $J = 1$ photon must therefore have the symmetry of the $\vec{B}$ field (magnetic dipole moment "$M - 1$" transition), and not of the $\vec{E}$ field (electric dipole moment "$E - 1$" transition). See Chap. 11 for more discussion of parity.

10. Symmetries

Problem 10.1 *Using isotopic spin decomposition for the decays of the ρ meson with $I = 1$: $\rho^+ \to \pi^+\pi^0$, $\rho^- \to \pi^-\pi^0$, $\rho^0 \to \pi^+\pi^-$ and $\rho^0 \to \pi^0\pi^0$, prove that $\rho^0 \to \pi^0\pi^0$ is forbidden on the basis of isospin invariance (that is, use the Adair–Shmushkevich analysis).*

The ρ as well as the π mesons belong to isotriplet states with $I = 1$. For ρ meson decaying into two π mesons, the possible reactions conserving all quantum numbers correspond to:

$$\begin{aligned} \rho^+ &\to \pi^+ + \pi^0, \\ \rho^0 &\to \pi^0 + \pi^0, \\ &\to \pi^+ + \pi^-, \\ \rho^- &\to \pi^- + \pi^0. \end{aligned} \tag{10.1}$$

Since isospin invariance for ρ meson states requires that the total decay rate for any charge state of ρ should add up to 1 (normalized), let us assume that

$$\text{Rate}(\rho^0 \to \pi^0 + \pi^0) = x, \quad \text{Rate}(\rho^0 \to \pi^+ + \pi^-) = 1 - x, \tag{10.2}$$

where x is yet to be determined. Thus, we can write a table of the form

Charge state of ρ	Final state	Rate
ρ^+	$\pi^+ + \pi^0$	1
ρ^0	$\pi^0 + \pi^0$	x
	$\pi^+ + \pi^-$	$1 - x$
ρ^-	$\pi^- + \pi^0$	1

Isospin invariance for the π meson in the final state would imply that

$$\begin{aligned}\text{Total Rate}(\pi^+ + \text{anything}) &= \text{Total Rate}(\pi^- + \text{anything})\\ &= \text{Total Rate}(\pi^0 + \text{anything}) \qquad (10.3)\\ \text{or} \qquad 1 + (1-x) &= (1-x) + 1 = 1 + 2x + 1,\end{aligned}$$

where, in the last expression, the two π^0 mesons in the final state of $\rho^0 \to \pi^0 + \pi^0$ are indistinguishable and both contribute to the count, which is the reason for the factor of 2. The solution to Eq. (10.3) yields

$$\begin{aligned}(1-x) + 1 &= 1 + 2x + 1\\ \text{or} \quad 3x = 0, &\quad \Rightarrow x = 0.\end{aligned} \qquad (10.4)$$

This implies that the decay

$$\rho^0 \to \pi^0 + \pi^0, \qquad (10.5)$$

is forbidden, while the other three decays take place at identical rates if isospin invariance holds.

The reason why the decay in (10.5) is forbidden can be understood from the composition of isospin quantum numbers (which is like angular momentum). The two π mesons in the final state belong to $I = 1$ states and therefore the isospin of the composite state can be

$$I = 2, 1 \text{ or } 0. \qquad (10.6)$$

The states with $I = 2$ and $I = 0$ are symmetric under exchange of two particles, while the $I = 1$ state is antisymmetric. This rules out a state containing two π^0 mesons from being in $I = 1$, because identical bosons must be in symmetric states. In fact, a state with two π^0 mesons can exist only in the $I = 2, 0$ states. (We will see this in greater detail in Problem 11.1 in the next chapter.) Thus, since $I(\rho) = 1$, this decay is forbidden if isospin is conserved in the process.

Problem 10.2 *Assuming invariance of strong interactions under rotations in isotopic-spin space and the usual isospin assignments for K and π mesons, what would you predict for the ratios of transition* *rates in the following decays:*

(a) *For an* $I = \frac{3}{2}$, K^* *meson,*

$$\frac{K^{*++} \to K^+\pi^+}{K^{*+} \to K^+\pi^0}, \qquad \frac{K^{*+} \to K^+\pi^0}{K^{*+} \to K^0\pi^+}, \qquad \frac{K^{*-} \to K^0\pi^-}{K^{*0} \to K^+\pi^-}.$$

(b) *What would you expect for the above processes if the* K^* *meson had* $I = \frac{1}{2}$? (*Hint: Consider the* I_3 *of the final states.*)

(a) Let us consider the two-body decay of the K^* meson

$$K^* \to K + \pi. \tag{10.7}$$

Since K^* mesons have a strangeness number $S(K^*) = 1$, this implies that

$$Y(K^*) = 1. \tag{10.8}$$

It then follows that if the K^* meson has $I = \frac{3}{2}$, there will be four charge states

$$Q(K^*) = I_3(K^*) + \frac{Y(K^*)}{2} = I_3(K^*) + \frac{1}{2} = 2, 1, 0, -1. \tag{10.9}$$

The possible decays of the K^* meson consistent with conservation of all quantum numbers can be written as

$$\begin{aligned}
K^{*++} &\to K^+ + \pi^+, \\
K^{*+} &\to K^+ + \pi^0, \\
&\to K^0 + \pi^+, \\
K^{*0} &\to K^+ + \pi^-, \\
&\to K^0 + \pi^0, \\
K^{*-} &\to K^0 + \pi^-,
\end{aligned} \tag{10.10}$$

where we have used the fact that

$$S(K^+) = S(K^0) = 1. \tag{10.11}$$

(For example, $K^{*0} \to K^- + \pi^+$ would violate strangeness conservation.)

If isospin invariance holds for the K^* meson, then the total decay rates for any charged state of K^* should add up to 1 (normalized). Thus, we can denote

$$\text{Rate}(K^{*+} \to K^+ + \pi^0) = x, \quad \text{Rate}(K^{*+} \to K^0 + \pi^+) = (1-x),$$
$$\text{Rate}(K^{*0} \to K^+ + \pi^-) = y, \quad \text{Rate}(K^{*0} \to K^0 + \pi^0) = (1-y), \tag{10.12}$$

where x, y are yet to be determined. We can therefore make the table

Charge state of K^*	Final state	Rate
K^{*++}	$K^+ + \pi^+$	1
K^{*+}	$K^+ + \pi^0$	x
	$K^0 + \pi^+$	$(1-x)$
K^{*0}	$K^+ + \pi^-$	y
	$K^0 + \pi^0$	$(1-y)$
K^{*-}	$K^0 + \pi^-$	1

Requiring isospin invariance for the K mesons in the final state, we obtain

$$\begin{aligned} &\text{Total Rate}(K^+ + \text{anything}) = \text{Total Rate}(K^0 + \text{anything}) \\ &\text{or} \quad 1 + x + y = (1-x) + (1-y) + 1. \end{aligned} \tag{10.13}$$

Similarly, isospin invariance for the π meson in the final state leads to

$$\begin{aligned} \text{Total Rate}(\pi^+ + \text{anything}) &= \text{Total Rate}(\pi^- + \text{anything}) \\ &= \text{Total Rate}(\pi^0 + \text{anything}) \\ \text{or} \quad 1 + (1-x) &= y + 1 = x + (1-y). \end{aligned} \tag{10.14}$$

Relationship (10.14) leads to

$$y = 1 - x, \tag{10.15}$$

which, in turn leads to

$$\begin{aligned} &1 + (1-x) = x + (1-y) = x + (1-(1-x)) = 2x \\ &\text{or} \quad 3x = 2 \\ &\text{or} \quad x = \frac{2}{3}, \quad y = 1 - x = 1 - \frac{2}{3} = \frac{1}{3}. \end{aligned} \tag{10.16}$$

Consequently, the relative rates become

$$\begin{aligned}
\frac{\text{Rate}(K^{*++} \to K^+ + \pi^+)}{\text{Rate}(K^{*+} \to K^+ + \pi^0)} &= \frac{1}{x} = \frac{1}{2/3} = \frac{3}{2},\\
\frac{\text{Rate}(K^{*+} \to K^+ + \pi^0)}{\text{Rate}(K^{*+} \to K^0 + \pi^+)} &= \frac{x}{1-x} = \frac{2/3}{1-2/3} = 2,\\
\frac{\text{Rate}(K^{*-} \to K^0 + \pi^-)}{\text{Rate}(K^{*0} \to K^+ + \pi^-)} &= \frac{1}{y} = \frac{1}{1/3} = 3.
\end{aligned} \tag{10.17}$$

(b) If the K^* has $I = \frac{1}{2}$, then there will be two charge states

$$Q(K^*) = I_3(K^*) + \frac{Y(K^*)}{2} = I_3(K^*) + \frac{1}{2} = 1, 0. \tag{10.18}$$

The possible decays consistent with all conservation laws are

$$\begin{aligned}
K^{*+} &\to K^+ + \pi^0,\\
&\to K^0 + \pi^+,\\
K^{*0} &\to K^+ + \pi^-,\\
&\to K^0 + \pi^0,
\end{aligned} \tag{10.19}$$

We can still parametrize the decay rates as in Eq. (10.12), and tabulate the results

Charge state of K^*	Final state	Rate
K^{*+}	$K^+ + \pi^0$	x
	$K^0 + \pi^+$	$(1-x)$
K^{*0}	$K^+ + \pi^-$	y
	$K^0 + \pi^0$	$(1-y)$

The requirement of isospin invariance for the K meson leads to

$$\begin{aligned}
\text{Total Rate}(K^+ + \text{anything}) &= \text{Total Rate}(K^0 + \text{anything})\\
\text{or} \quad x + y &= (1-x) + (1-y).
\end{aligned} \tag{10.20}$$

Similarly, isospin invariance for the π meson states leads to

$$\begin{aligned}
\text{Total Rate}(\pi^+ + \text{anything}) &= \text{Total Rate}(\pi^- + \text{anything})\\
&= \text{Total Rate}(\pi^0 + \text{anything})\\
\text{or} \quad (1-x) = y &= x + (1-y).
\end{aligned} \tag{10.21}$$

Substituting $y = 1 - x$ from the above relationship in Eq. (10.21), and solving for x, we obtain

$$\begin{aligned} 1 - x &= x + (1 - y) = x + (1 - (1 - x)) = 2x \\ \text{or}\quad 3x &= 1 \\ \text{or}\quad x &= \frac{1}{3}, \quad y = 1 - x = 1 - \frac{1}{3} = \frac{2}{3}. \end{aligned} \tag{10.22}$$

Calculating relative rates, we obtain

$$\begin{aligned} \frac{\text{Rate}(K^{*+} \to K^+ + \pi^0)}{\text{Rate}(K^{*+} \to K^0 + \pi^+)} &= \frac{x}{1-x} = \frac{1/3}{1-1/3} = \frac{1}{2}, \\ \frac{\text{Rate}(K^{*+} \to K^+ + \pi^0)}{\text{Rate}(K^{*0} \to K^+ + \pi^-)} &= \frac{x}{y} = \frac{1/3}{2/3} = \frac{1}{2}, \\ \frac{\text{Rate}(K^{*0} \to K^+ + \pi^-)}{\text{Rate}(K^{*0} \to K^0 + \pi^0)} &= \frac{y}{1-y} = \frac{2/3}{1-2/3} = 2. \end{aligned} \tag{10.23}$$

Such relative rates can be measured, and thereby used to determine the isospin of any decaying particle.

Problem 10.3 *N^* baryons are $I = \frac{1}{2}$ excited states of the nucleon. On the basis of isospin invariance in strong interactions, compare the differences expected for N^* and Δ decays into the π–N systems discussed in Table* 10.2.

The Δ particle has $I = \frac{3}{2}$ and has two-body decays of the form

$$\Delta \to N + \pi, \tag{10.24}$$

where N denotes collectively p or n, which belong to an $I = \frac{1}{2}$ doublet. Thus, these decay processes are identical to those handled in Problem 10.2(a), but now involving nonstrange hadrons. Thus, with the following identification

$$\Delta \leftrightarrow K^*, \quad p \leftrightarrow K^+, \quad n \leftrightarrow K^0, \tag{10.25}$$

if isospin invariance holds, the calculation is identical to Problem 10.2(a), and the results are the same as those given in Table 10.2 of the text.

The N^* baryons, on the other hand, are excited states of nucleons N, with $I = \frac{1}{2}$, and have two-body decays of the form

$$N^* \to N + \pi. \tag{10.26}$$

These decay processes are again completely parallel to those of Problem 10.2(b) (but involve nonstrange baryons), with the correspondence

$$N^* \leftrightarrow K^* \left(I = \frac{1}{2}\right), \quad p \leftrightarrow K^+, \quad n \leftrightarrow K^0. \tag{10.27}$$

Hence, if isospin invariance holds, the results for the decays of N^* can be obtained directly from Problem 10.2(b).

Problem 10.4 *What are the possible values of isotopic spin for the following systems?* (a) *A* π^+ *meson and an antiproton,* (b) *two neutrons,* (c) *a* π^+ *meson and a* Λ^0, (d) *a* π^+ *and a* π^0 *meson,* (e) *a* u *and a* $\bar{u}$ *quark,* (f) *a* c, b *and an* s *quark (for properties of quarks, see Table* 9.5).

Strong isotopic spins for composite systems can be determined in the following manner. Isotopic spins for several hadrons are listed in Table 9.3 of the text, and we derived in Problem 9.5 the isotopic spins of quarks: $I(u) = \frac{1}{2} = I(d)$ while $I(s) = I(c) = I(b) = I(t) = 0$.

(a) Using the composition law for isospin, we get

$$\begin{aligned} I(\pi^+\bar{p}) &= \frac{3}{2}, \frac{1}{2}, \\ I_3(\pi^+\bar{p}) &= I_3(\pi^+) + I_3(\bar{p}) = 1 - \frac{1}{2} = \frac{1}{2}. \end{aligned} \tag{10.28}$$

Since $I_3 = \frac{1}{2}$ can be a legitimate isospin projection for both $I = \frac{3}{2}$ and $I = \frac{1}{2}$, we conclude that the composite state $(\pi^+\bar{p})$ can have either $I = \frac{3}{2}$ or $I = \frac{1}{2}$.

(b) A composite state of two neutrons can have

$$\begin{aligned} I(nn) &= 1, 0 \\ I_3(nn) &= I_3(n) + I_3(n) = -\frac{1}{2} - \frac{1}{2} = -1. \end{aligned} \tag{10.29}$$

Since $I_3 = -1$ cannot be an isospin projection of a state with $I = 0$, we conclude that a composite state of two neutrons can only be $I = 1$.

(c) The Λ^0 particle is an isosinglet ($I = 0$) and, consequently, we have

$$I(\pi^+\Lambda^0) = 1. \tag{10.30}$$

This composite state is therefore $I = 1$.

(d) We can have

$$\begin{aligned} I(\pi^+\pi^0) &= 2, 1, 0 \\ I_3(\pi^+\pi^0) &= I_3(\pi^+) + I_3(\pi^0) = 1 + 0 = 1. \end{aligned} \tag{10.31}$$

Thus, as we have already discussed in Problem 10.1, a state with two π mesons can have $I = 2$, $I = 1$ or $I = 0$. However, $I = 0$ has no projection of $I_3 = 1$, and the composite state $(\pi^+\pi^0)$ must therefore have $I = 2$ or $I = 1$.

(e) For a composite system of a u and a $\bar{u}$ quark, we have

$$\begin{aligned} I(u\bar{u}) &= 1, 0, \\ I_3(u\bar{u}) &= I_3(u) + I_3(\bar{u}) = \frac{1}{2} - \frac{1}{2} = 0. \end{aligned} \tag{10.32}$$

Since the projection $I_3 = 0$ is possible for both $I = 1, 0$ states, we conclude that the composite state $(u\bar{u})$ can have $I = 1$ or $I = 0$.

(f) As we have seen in Problem 9.5, all quarks with flavor quantum numbers are isosinglets. Thus, we have

$$I(bcs) = 0. \tag{10.33}$$

This composite state would therefore be a baryon with unique isospin $I = 0$.

11. Discrete Transformations

Problem 11.1 *The $\rho^0(770)$ has $J^P = 1^-$, and it decays strongly into $\pi^+\pi^-$ pairs. From symmetry and angular momentum considerations, explain why the decay $\rho^0(770) \to \pi^0\pi^0$ is forbidden.*

As we already indicated briefly in Problem 10.1, isospin invariance forbids the decay

$$\rho^0(770) \to \pi^0 + \pi^0. \tag{11.1}$$

Let us analyze this in some more detail.

We know that the ρ^0 meson has $I = 1, I_3 = 0$. We have also noted in Problem 10.1 that a state with two π mesons can have isospin $I = 2, 1$ or 0. Using the composition law of angular momentum, we can construct these isospin states explicitly. The five projection states of the $I = 2$ state are symmetric under the exchange of the two π mesons and have the normalized forms:

$$\begin{aligned}
|I = 2, I_3 = 2\rangle &= |\pi^+\pi^+\rangle, \\
|I = 2, I_3 = 1\rangle &= \frac{1}{\sqrt{2}}(|\pi^+\pi^0\rangle + |\pi^0\pi^+\rangle), \\
|I = 2, I_3 = 0\rangle &= \frac{1}{\sqrt{6}}(2|\pi^0\pi^0\rangle + |\pi^+\pi^-\rangle + |\pi^-\pi^+\rangle), \\
|I = 2, I_3 = -1\rangle &= \frac{1}{\sqrt{2}}(|\pi^0\pi^-\rangle + |\pi^-\pi^0\rangle), \\
|I = 2, I_3 = -2\rangle &= |\pi^-\pi^-\rangle.
\end{aligned} \tag{11.2}$$

On the other hand, the three projection states of $I = 1$ are anti-symmetric under the exchange of the two π mesons. They have the explicit normalized forms

$$|I=1,I_3=1\rangle = \frac{1}{\sqrt{2}}(|\pi^+\pi^0\rangle - |\pi^0\pi^+\rangle),$$

$$|I=1,I_3=0\rangle = \frac{1}{\sqrt{2}}(|\pi^+\pi^-\rangle - |\pi^-\pi^+\rangle), \qquad (11.3)$$

$$|I=1,I_3=-1\rangle = \frac{1}{\sqrt{2}}(|\pi^0\pi^-\rangle - |\pi^-\pi^0\rangle).$$

Finally, the isosinglet state is symmetric under the exchange of the two pions and has the normalized form

$$|I=0,I_3=0\rangle = \frac{1}{\sqrt{3}}(-|\pi^0\pi^0\rangle + |\pi^+\pi^-\rangle + |\pi^-\pi^+\rangle). \qquad (11.4)$$

Thus, we see that the $I=1, I_3=0$ state (corresponding to the isospin quantum numbers of the ρ^0 meson) does not have a $|\pi^0\pi^0\rangle$ component simply because the required antisymmetry of the state does not allow this. As a result, the decay (11.1) cannot take place if isospin invariance holds. As mentioned in Problem 10.1, only the symmetric states with $I=2,0$ and $I_3=0$ have $|\pi^0\pi^0\rangle$ components. From a less formal perspective, because in the rest frame of the ρ, the two π^0 mesons are identical and indistinguishable, both their isospin and their orbital angular momentum states must be symmetric. However, the ρ has $J^P=1^-$ and $J^P_\pi=0$, and the two pions must consequently be in the $\ell=1$ to conserve angular momentum. But this is an odd state of the system, and therefore forbidden for two π^0 mesons.

Problem 11.2 *What is the charge-conjugate reaction to* $K^- + p \to \bar{K}^0 + n$*? Can a* K^-p *system be an eigenstate of the charge conjugation operator? Similarly, discuss the reaction* $\bar{p}+p \to \pi^+ + \pi^-$*.*

Under charge conjugation, we have

$$C|K^-\rangle = |K^+\rangle, \quad C|p\rangle = |\bar{p}\rangle, \quad C|n\rangle = |\bar{n}\rangle, \quad C|K^0\rangle = |\bar{K}^0\rangle. \qquad (11.5)$$

Since $C^2=1$, it follows that

$$C|K^+\rangle = |K^-\rangle, \quad C|\bar{K}^0\rangle = |K^0\rangle, \quad C|\bar{p}\rangle = |p\rangle, \qquad (11.6)$$

and we see that

$$C|K^-p\rangle = |K^+\bar{p}\rangle, \quad C|\bar{K}^0n\rangle = |K^0\bar{n}\rangle. \tag{11.7}$$

The charge conjugate of the reaction

$$K^- + p \to \bar{K}^0 + n \tag{11.8}$$

is therefore

$$K^+ + \bar{p} \to K^0 + \bar{n}. \tag{11.9}$$

Although the state $|K^-p\rangle$ is charge neutral (which is a necessary condition to be an eigenstate of C, as discussed after Eq. (11.54) of the text), from Eq. (11.7) we see that $|K^+\bar{p}\rangle$ represents a physically distinct state from $|K^-p\rangle$. Consequently, the state $|K^-p\rangle$ is not an eigenstate of the charge conjugation operator. (Recall that a state $|\psi\rangle$ is an eigenstate of C if it satisfies $C|\psi\rangle = \eta_C|\psi\rangle$, with $\eta_C^2 = 1$.)

Since we know that

$$C|\pi^+\rangle = |\pi^-\rangle, \quad C|\pi^-\rangle = |\pi^+\rangle, \quad C|\pi^0\rangle = |\pi^0\rangle, \tag{11.10}$$

for the reaction

$$\bar{p} + p \to \pi^+ + \pi^-, \tag{11.11}$$

the charge conjugate reaction is given by

$$p + \bar{p} \to \pi^- + \pi^+, \tag{11.12}$$

which is the same as the original reaction. Hence, under charge conjugation, we get

$$C|\bar{p}p\rangle = |p\bar{p}\rangle, \quad C|\pi^+\pi^-\rangle = |\pi^-\pi^+\rangle. \tag{11.13}$$

We can form symmetric and antisymmetric combinations of these states that are eigenstates of the charge conjugation operator, namely ($\frac{1}{\sqrt{2}}$ is just a normalization factor):

$$\begin{aligned} C\left(\frac{1}{\sqrt{2}}\left(|\bar{p}p\rangle \pm |p\bar{p}\rangle\right)\right) &= \pm\frac{1}{\sqrt{2}}\left(|\bar{p}p\rangle \pm |p\bar{p}\rangle\right), \\ C\left(\frac{1}{\sqrt{2}}(|\pi^+\pi^-\rangle \pm |\pi^-\pi^+\rangle)\right) &= \pm\frac{1}{\sqrt{2}}(|\pi^+\pi^-\rangle \pm |\pi^-\pi^+\rangle). \end{aligned} \tag{11.14}$$

A proton–antiproton system can therefore be in an eigenstate of C, just as a π^+ and a π^- system. Reaction (11.11) can therefore proceed, preserving charge conjugation either for $C = +$ or $C = -$ states. (Again, the reverse reaction can only proceed virtually.)

Problem 11.3 *If ρ^0 mesons are produced in states with spin projection $J_z = 0$ along their line of flight, what would you expect for the angular distribution of $\rho^0 \to \pi^+ + \pi^-$ decay products in the ρ^0 rest frame? (See Appendix B for the appropriate $Y_{\ell,m}(\theta,\phi)$ functions.) What would be your answer if the initial ρ^0 had spin projection $J_z = +1$?*

Let us assume that the line of flight of the produced ρ^0 meson defines the z-axis. In the rest frame of the ρ^0 meson, the π^+ and π^- mesons will be produced back-to-back in order to conserve momentum.

The spin-parity quantum numbers in the strong decay

$$\rho^0 \to \pi^+ + \pi^-, \tag{11.15}$$

are

$$1^- \to 0^- + 0^-. \tag{11.16}$$

Thus, we see that conservation of angular momentum (as well as parity, since it is a strong process) requires that the relative orbital angular momentum of the final state be

$$\ell = 1. \tag{11.17}$$

Because pions are spin-zero particles, this represents the total angular momentum of the final state.

(a) If the initial ρ^0 meson has $J_z = 0$, then conservation of angular momentum (projection) requires that the final state of two pions must have

$$\ell_z = m = 0. \tag{11.18}$$

The spatial component of the wave function will have the form

$$\psi_{\pi^+\pi^-}(r,\theta,\phi) \approx f(r)Y_{10}(\theta,\phi) \approx f(r)\cos\theta, \tag{11.19}$$

where $f(r)$ is the radial component of the wave function, which is not relevant for our discussion. The form of the spherical harmonics is from Eq. (B.6) of the text. (The isospin component of this decay has already been discussed in Problem 11.1.) The angular distribution of the decay products is therefore

$$\frac{d\sigma}{d\Omega} \sim |\psi_{\pi^+\pi^-}(r,\theta,\phi)|^2 \sim |Y_{10}(\theta,\phi)|^2 \sim \cos^2\theta. \tag{11.20}$$

(b) If the decaying ρ^0 meson is in the spin state $J_z = \pm 1$, then conservation of angular momentum implies that the final state should have

$$\ell_z = m = \pm 1. \tag{11.21}$$

The final state wave function will have the form

$$\psi_{\pi^+\pi^-}(r,\theta,\phi) = f(r)Y_{1,\pm 1}(\theta,\phi) \sim f(r)\sin\theta\; e^{\pm i\phi}. \tag{11.22}$$

This leads to an angular distribution of the form

$$\frac{d\sigma}{d\Omega} \sim |\psi_{\pi^+\pi^-}(r,\theta,\phi)|^2 \sim |Y_{1,\pm 1}(\theta,\phi)|^2 \sim \sin^2\theta. \tag{11.23}$$

The angular distributions in Eqs. (11.20) and (11.23) can be distinguished in data to provide the J_z of the original ρ meson, and thereby yield information on its production mechanism.

Problem 11.4 *The Ξ^- has $J^P = \frac{1}{2}^+$. It decays through weak interaction into a Λ^0 and a π^- meson. If $J^P_\Lambda = \frac{1}{2}^+$ and $J^P_\pi = 0^-$, what are the allowed relative orbital angular momenta for the Λ–π^- system?*

The reaction

$$\Xi^- \to \Lambda^0 + \pi^-, \tag{11.24}$$

is a weak decay process. We therefore expect several quantum numbers to be violated. Looking at Table 9.4 of the text, it is clear that this process violates both isospin and strangeness quantum numbers, as is the case in many weak hadronic decays. The spin-parity assignments in the reaction are:

$$\frac{1}{2}^{+} \to \frac{1}{2}^{+} + 0^{-}. \tag{11.25}$$

If we assume the final state particles have a relative orbital angular momentum $\ell_{\Lambda^0\pi^-}$, then from conservation of total angular momentum we obtain

$$\begin{aligned} J_{\Xi^-} &= J_{\Lambda^0\pi^-} \\ \text{or} \quad \frac{1}{2} &= \ell_{\Lambda^0\pi^-} + \frac{1}{2}, \quad \text{or} \quad \left|\ell_{\Lambda^0\pi^-} - \frac{1}{2}\right|. \end{aligned} \tag{11.26}$$

This determines

$$\ell_{\Lambda^0\pi^-} = 0, 1. \tag{11.27}$$

If $\ell_{\Lambda^0\pi^-} = 0$, then we see from (11.25) that parity will be violated in this process (see Eq. (11.21) of the text). On the other hand, if $\ell_{\Lambda^0\pi^-} = 1$, parity will be conserved in the decay. From the discussion in Sec. 11.2.2 of the text, we know that parity can be violated in weak interactions, and both these orbital angular momentum values are therefore allowed in the decay.

Problem 11.5 *Which of the following decays are forbidden by C-invariance?*

(a) $\omega^0 \to \pi^0 + \gamma$,
(b) $\eta' \to \rho^0 + \gamma$,
(c) $\pi^0 \to \gamma + \gamma + \gamma$,
(d) $J/\psi \to \bar{p} + p$,
(e) $\rho^0 \to \gamma + \gamma$.

(*Check the CRC tables to see if these decays take place.*)

From the *CRC Handbook* or from the *PDG*, we have

$$\begin{aligned} &\eta_C(\gamma) = -1, \quad \eta_C(\pi^0) = 1, \quad \eta_C(\rho^0) = -1, \\ &\eta_C(\omega^0) = -1, \quad \eta_C(J/\psi) = -1, \quad \eta_C(\eta') = 1, \end{aligned} \tag{11.28}$$

where η_C represents the charge parity of the state. Now we can check whether charge parity is conserved in the processes.

(a) In the reaction

$$\omega \rightarrow \pi^0 + \gamma, \tag{11.29}$$

the initial state has the charge parity

$$\eta_C^{(\text{initial})} = \eta_C(\omega^0) = -1, \tag{11.30}$$

whereas the final state has the charge parity

$$\eta_C^{(\text{final})} = \eta_C(\pi^0)\eta_C(\gamma) = 1 \times (-1) = -1. \tag{11.31}$$

Thus, we see that charge parity is conserved in the process (11.29) and it is therefore allowed by C invariance.

(b) The initial state charge parity of the reaction

$$\eta' \rightarrow \rho^0 + \gamma, \tag{11.32}$$

is given by

$$\eta_C^{(\text{initial})} = \eta_C(\eta') = 1. \tag{11.33}$$

The final state has the charge parity

$$\eta_C^{(\text{final})} = \eta_C(\rho^0)\eta_C(\gamma) = (-1) \times (-1) = 1. \tag{11.34}$$

Thus, charge parity is conserved in this process, and it is allowed by C invariance.

(c) From the general result of Eq. (11.60) of the text, we know that a π^0 meson cannot decay into an odd number of photons if C invariance holds. This can be checked explicitly in the present case

$$\pi^0 \rightarrow \gamma + \gamma + \gamma, \tag{11.35}$$

for which

$$\eta_C^{(\text{initial})} = \eta_C(\pi^0) = 1, \quad \eta_C^{(\text{final})} = (\eta_C(\gamma))^3 = (-1)^3 = -1. \tag{11.36}$$

Thus, this reaction is not allowed by C invariance.

(d) For the process

$$J/\psi \rightarrow p + \bar{p}, \tag{11.37}$$

we note that the charge parity of the initial state is

$$\eta_C(J/\psi) = -1. \tag{11.38}$$

The individual particles in the final state $(p, \bar{p})$ are charged and therefore cannot correspond to eigenstates of the charge conjugation operator. However, as we have already seen in Problem 11.2

(see Eq. (11.14)), we can have eigenstates of C comprised of a p and a $\bar{p}$ (the composite state is charge neutral):

$$C\left(\frac{1}{\sqrt{2}}\left(|\bar{p}p\rangle \pm |p\bar{p}\rangle\right)\right) = \pm\frac{1}{\sqrt{2}}\left(|\bar{p}p\rangle \pm |p\bar{p}\rangle\right). \tag{11.39}$$

The charge parity of the antisymmetric state is given by

$$\eta_C((p\bar{p})_{\text{anti}}) = -1, \tag{11.40}$$

which agrees with that for the initial state. Therefore, the J/ψ can preserve C invariance when it decays into an odd-C state. $|p\bar{p}\rangle$ is in a charge odd state.

(e) For the decay

$$\rho^0 \to \gamma + \gamma, \tag{11.41}$$

we note that

$$\eta_C^{(\text{initial})} = \eta_C(\rho^0) = -1, \quad \eta_C^{(\text{final})} = (\eta_C(\gamma))^2 = (-1)^2 = 1. \tag{11.42}$$

Therefore, this decay is not allowed by C invariance.

Problem 11.6 *Although the orbital wave for any strong π–N state determines the parity of that state, different ℓ-values do not necessarily yield different decay angular distributions. In particular, show that a $J = \frac{1}{2}$, $J_Z = +\frac{1}{2}$, π–N resonance decays the same way whether it has $\ell = 0$ or $\ell = 1$. Similarly, show that a $J = \frac{3}{2}$, $J_Z = +\frac{1}{2}$, π–N system has the same decay angular distribution for $\ell = 1$ as for $\ell = 2$. [Hint: Expand the wave function for the state in terms of the products of $s = \frac{1}{2}$ spin-states and the appropriate $Y_{\ell,m}(\theta,\phi)$.]*

For a $\pi - N$ system, from the relationship

$$\vec{J}_{\pi N} = \vec{L}_{\pi N} + \vec{S}_N \tag{11.43}$$

the possible values of the total angular momentum are

$$j = \ell + \frac{1}{2}, \ \left|\ell - \frac{1}{2}\right|,$$
$$m_j = m_\ell + m_s = m_\ell \pm \frac{1}{2}, \tag{11.44}$$

where we denote by $j, \ell, m_j, m_\ell, m_s$ the eigenvalues of the total angular momentum, relative orbital angular momentum, the projections of total angular momentum, relative orbital angular momentum, and intrinsic-spin angular momentum of the nucleon, respectively.

(a) If we have $j = \frac{1}{2}$ and $m_j = \frac{1}{2}$ for the $\pi - N$ system (eigenvalues of J, J_z), then from Eq. (11.44) we see that the allowed values of relative $\pi - N$ orbital angular momentum are given by

$$\ell = 0, 1. \tag{11.45}$$

For $\ell = 0$, we have $j = \frac{1}{2}$, while $\ell = 1$ leads to $j = \frac{3}{2}$ and $\frac{1}{2}$. The eigenstates $|j, m_j\rangle$ can be constructed in terms of the basis states $|\ell; m_\ell, m_s\rangle$ using the Adair–Shmushkevitch formalism developed in the text, or the more standard composition law of angular momentum. Now, let us write the eigenstates of $j = \frac{1}{2}$ corresponding to $\ell = 0$ and $\ell = 1$ (the states can have an arbitrary overall phase):

$$\begin{aligned}
\left|j = \frac{1}{2}, m_j = \frac{1}{2}\right\rangle^{(\ell=1)} &= \frac{1}{\sqrt{3}}\left(\left|m_\ell = 0, m_s = \frac{1}{2}\right\rangle \right. \\
&\quad \left. - \sqrt{2}\left|m_\ell = 1, m_s = -\frac{1}{2}\right\rangle\right), \\
\left|j = \frac{1}{2}, m_j = -\frac{1}{2}\right\rangle^{(\ell=1)} &= \frac{1}{\sqrt{3}}\left(\sqrt{2}\left|m_\ell = -1, m_s = \frac{1}{2}\right\rangle \right. \\
&\quad \left. - \left|m_\ell = 0, m_s = -\frac{1}{2}\right\rangle\right), \\
\left|j = \frac{1}{2}, m_j = \frac{1}{2}\right\rangle^{(\ell=0)} &= \left|m_\ell = 0, m_s = \frac{1}{2}\right\rangle, \\
\left|j = \frac{1}{2}, m_j = -\frac{1}{2}\right\rangle^{(\ell=0)} &= \left|m_\ell = 0, m_s = -\frac{1}{2}\right\rangle.
\end{aligned} \tag{11.46}$$

Ignoring the radial part of the wave functions, which are not relevant to our discussion, we note that the angular parts of the system with $j = \frac{1}{2} = m_j$, corresponding to $\ell = 0$ or $\ell = 1$ can be written as

$$\psi^{(\ell=0)}_{j=1/2,m_j=1/2}(r,\theta,\phi) \approx Y_{00}(\theta,\phi)\left|m_s=\frac{1}{2}\right\rangle = \frac{1}{\sqrt{4\pi}}\left|m_s=\frac{1}{2}\right\rangle,$$

$$\begin{aligned}\psi^{(\ell=1)}_{j=1/2,m_j=1/2}&(r,\theta,\phi)\\ &\approx \frac{1}{\sqrt{3}}\left(Y_{10}(\theta,\phi)\left|m_s=\frac{1}{2}\right\rangle - \sqrt{2}Y_{11}(\theta,\phi)\left|m_s=-\frac{1}{2}\right\rangle\right)\\ &= \frac{1}{\sqrt{4\pi}}\left(\cos\theta\left|m_s=\frac{1}{2}\right\rangle - \sin\theta\, e^{i\phi}\left|m_s=-\frac{1}{2}\right\rangle\right),\end{aligned} \tag{11.47}$$

where ϕ denotes the azimuthal angle, and we have used the forms of the spherical harmonics given in Eq. (B.6) of the text. It follows that the angular parts of the square of the wave functions have the forms:

$$\begin{aligned}|\psi^{(\ell=0)}_{j=1/2,m_j=1/2}|^2 &\approx \frac{1}{4\pi}\left\langle m_s=\frac{1}{2}\middle|m_s=\frac{1}{2}\right\rangle = \frac{1}{4\pi},\\ |\psi^{(\ell=1)}_{j=1/2,m_j=1/2}|^2 &\approx \frac{1}{4\pi}\left(\cos^2\theta\left\langle m_s=\frac{1}{2}\middle|m_s=\frac{1}{2}\right\rangle\right.\\ &\qquad\left.+\sin^2\theta\left\langle m_s=-\frac{1}{2}\middle|m_s=-\frac{1}{2}\right\rangle\right)\\ &= \frac{1}{4\pi}\left(\cos^2\theta+\sin^2\theta\right) = \frac{1}{4\pi}.\end{aligned} \tag{11.48}$$

The cross terms in the above expression vanish because of the orthogonality of the spin wave functions, and we see that the angular distributions in both cases are isotropic. (The same would hold for the orthogonal $m_j = -\frac{1}{2}$.) Because we are only interested in the angular part of the solution, we have again suppressed the radial dependence of the wave function, which is not relevant for our discussion.

(b) If the $\pi - N$ system is in a state with $j = \frac{3}{2}, m_j = \frac{1}{2}$, the allowed values of relative orbital angular momentum are $\ell = 2, 1$. If $\ell = 2$, we can have $j = \frac{5}{2}, \frac{3}{2}$, while $\ell = 1$ will yield $j = \frac{3}{2}, \frac{1}{2}$. Once again, all the states can be constructed. For simplicity, we record only the relevant states for the projection $m_j = 1/2$

$$\begin{aligned}
\left|j=\frac{3}{2},m_j=\frac{1}{2}\right\rangle^{(\ell=1)} &= \frac{1}{\sqrt{3}}\left(\left|m_\ell=0,m_s=\frac{1}{2}\right\rangle\right. \\
&\quad \left.+\left|m_\ell=1,m_s=-\frac{1}{2}\right\rangle\right), \\
\left|j=\frac{3}{2},m_j=\frac{1}{2}\right\rangle^{(\ell=2)} &= \frac{1}{\sqrt{5}}\left(\sqrt{2}\left|m_\ell=0,m_s=\frac{1}{2}\right\rangle\right. \\
&\quad \left.-\sqrt{3}\left|m_\ell=1,m_s=-\frac{1}{2}\right\rangle\right).
\end{aligned} \tag{11.49}$$

Again disregarding the radial part of the wave functions, the angular and the spin parts are:

$$\begin{aligned}
\psi^{(\ell=1)}_{j=3/2,m_j=1/2} &\approx \frac{1}{\sqrt{3}}\left(\sqrt{2}Y_{10}(\theta,\phi)\left|m_s=\frac{1}{2}\right\rangle+Y_{11}(\theta,\phi)\left|m_s=-\frac{1}{2}\right\rangle\right) \\
&= \frac{1}{\sqrt{8\pi}}\left(2\cos\theta\left|m_s=\frac{1}{2}\right\rangle-\sin\theta\; e^{i\phi}\left|m_s=-\frac{1}{2}\right\rangle\right), \\
\psi^{(\ell=2)}_{j=3/2,m_j=1/2} &\approx \frac{1}{\sqrt{5}}\left(\sqrt{2}Y_{20}(\theta,\phi)\left|m_s=\frac{1}{2}\right\rangle\right. \\
&\quad \left.-\sqrt{3}Y_{21}(\theta,\phi)\left|m_s=-\frac{1}{2}\right\rangle\right) \\
&= \frac{1}{\sqrt{8\pi}}\left((3\cos^2\theta-1)\left|m_s=\frac{1}{2}\right\rangle\right. \\
&\quad \left.+3\sin\theta\cos\theta\; e^{i\phi}\left|m_s=-\frac{1}{2}\right\rangle\right).
\end{aligned} \tag{11.50}$$

Taking the absolute squares of the wave functions yield

$$\begin{aligned}
|\psi^{(\ell=1)}_{j=3/2,m_j=1/2}|^2 &\approx \frac{1}{8\pi}\left(4\cos^2\theta\left\langle m_s=\frac{1}{2}\middle|m_s=\frac{1}{2}\right\rangle\right. \\
&\quad \left.+\sin^2\theta\left\langle m_s=-\frac{1}{2}\middle|m_s=-\frac{1}{2}\right\rangle\right) \\
&= \frac{1}{8\pi}(4\cos^2\theta+\sin^2\theta) = \frac{1}{8\pi}(1+3\cos^2\theta),
\end{aligned}$$

$$|\psi_{j=3/2,m_j=1/2}^{(\ell=2)}|^2 \approx \frac{1}{8\pi}\left((3\cos^2\theta - 1)^2\left\langle m_s = \frac{1}{2}\middle| m_s = \frac{1}{2}\right\rangle\right.$$

$$\left. + \; 9\sin^2\theta\cos^2\theta\left\langle m_s = -\frac{1}{2}\middle| m_s = -\frac{1}{2}\right\rangle\right)$$

$$= \frac{1}{8\pi}\left((3\cos^2\theta - 1)^2 + 9\sin 2\theta\cos^2\theta\right)$$

$$= \frac{1}{8\pi}\left(1 + 3\cos^2\theta\right). \tag{11.51}$$

We see once again that the angular distributions for both orbital values (but opposite parities) are the same. (And again, the same result would hold for $m_j = -\frac{1}{2}$.)

12. CP Violation

Problem 12.1 *Ignoring CP violation, plot to* $\approx 10\%$ *accuracy the probability of observing of* $\bar{K}^0$ *as a function of time in a beam that is initially* $(t = 0)$ *pure* K^0.

The probability of observing $\bar{K}^0$ as a function of time in a beam that is initially pure K^0 is given in Eq. (12.58) of the text. In that chapter, for simplicity, we had set $c = 1$, but here we re-introduce the factor of c to make the numerical calculations more transparent:

$$P(\bar{K}^0, t) = \frac{1}{4}\left|\frac{q}{p}\right|^2 \left[e^{-\frac{t}{\tau_S}} + e^{-\frac{t}{\tau_L}} - 2e^{-\frac{1}{2}\left(\frac{1}{\tau_S}+\frac{1}{\tau_L}\right)t} \cos\frac{\Delta mc^2}{\hbar}t\right], \tag{12.1}$$

where

$$\begin{aligned}
\tau_S &\approx 0.9 \times 10^{-10}\,\text{sec}, \quad \tau_L \approx 5 \times 10^{-8}\,\text{sec}, \\
\Delta m &= m_L - m_S \approx 3.5 \times 10^{-12}\,\text{MeV}/c^2, \\
p &= 1 + \epsilon, \quad q = -1 + \epsilon,
\end{aligned} \tag{12.2}$$

with ϵ representing the parameter of CP violation. Now,

$$\begin{aligned}
\frac{\Delta mc^2}{\hbar} &= \frac{\Delta mc^2}{\hbar c} \times c \approx \frac{3.5 \times 10^{-12}\,\text{MeV}}{200\,\text{MeV} - \text{F}} \times 3 \times 10^{10}\,\text{cm/sec} \\
&\approx 5.25 \times 10^{9}/\text{sec} \approx \frac{1}{2\tau_S},
\end{aligned} \tag{12.3}$$

$$\frac{\tau_S}{\tau_L} \approx \frac{0.9 \times 10^{-10}\,\text{sec}}{5 \times 10^{-8}\,\text{sec}} \approx 2 \times 10^{-3}.$$

From Eq. (12.42) of the text, we see that in the absence of the small CP violation $(\epsilon = 0)$, the weak eigenstates $|K_S^0\rangle, |K_L^0\rangle$ coincide with the strong eigenstates $|K_1^0\rangle, |K_2^0\rangle$ of CP. Thus, ignoring CP violation,

$$p = -q, \quad \left|\frac{q}{p}\right|^2 = 1, \tag{12.4}$$

so that

$$\begin{aligned} P(\bar{K}^0, t) &= \frac{1}{4}\left[e^{-\frac{t}{\tau_S}} + e^{-\frac{t}{\tau_L}} - 2e^{-\frac{1}{2}\left(\frac{1}{\tau_S}+\frac{1}{\tau_L}\right)t} \cos\frac{\Delta mc^2}{\hbar}t\right] \\ &\approx \frac{1}{4}\left[e^{-\frac{t}{\tau_S}} + e^{-\frac{t}{\tau_L}} - 2e^{-\frac{1}{2}\left(\frac{1}{\tau_S}+\frac{1}{\tau_L}\right)t} \cos\frac{t}{2\tau_S}\right]. \end{aligned} \tag{12.5}$$

Comparing with Eq. (12.1), we see that the qualitative behavior of the probability is the same with or without CP violation, since the difference lies in an overall multiplicative factor $\left|\frac{q}{p}\right|^2$. The probability in (12.5) clearly vanishes at $t = 0$, as it should, since the initial beam is pure K^0. For $\tau_L \gg \tau_S > t$, the probability for observing $\bar{K}^0$ increases quadratically with t, as can be seen from the fact that, for $t = 0$, Eq. (12.5) leads to

$$\begin{aligned} P(\bar{K}^0, t) &\approx \frac{1}{4}\Bigg[1 - \frac{t}{\tau_S} + \frac{t^2}{8\tau_S^2} + 1 - \frac{t}{\tau_L} + \frac{t^2}{8\tau_L^2} \\ &\quad -2\left(1 - \frac{t}{2}\left(\frac{1}{\tau_S} + \frac{1}{\tau_L}\right) + \frac{t^2}{8}\left(\frac{1}{\tau_S} + \frac{1}{\tau_L}\right)^2\right)\left(1 - \frac{t^2}{8\tau_S^2}\right)\Bigg] \\ &\approx \frac{1}{4}\left[2 - \frac{t}{\tau_S} + \frac{t^2}{2\tau_S^2} - 2\left(1 - \frac{t}{2\tau_S} + \frac{t^2}{8\tau_S^2}\right)\left(1 - \frac{t^2}{8\tau_S^2}\right)\right] \\ &\approx \frac{t^2}{8\tau_S^2}. \end{aligned} \tag{12.6}$$

For $\tau_L \gg t > \tau_S$, the probability in Eq. (12.5) can be written approximately as

$$\begin{aligned} P(\bar{K}^0, t) &\approx \frac{1}{4}\left[e^{-\frac{t}{\tau_L}} - 2e^{-\frac{t}{2\tau_S}} \cos\frac{t}{2\tau_s}\right] \\ &\approx \frac{1}{4}\left[1 - 2e^{-\frac{t}{2\tau_S}} \cos\frac{t}{2\tau_S}\right]. \end{aligned} \tag{12.7}$$

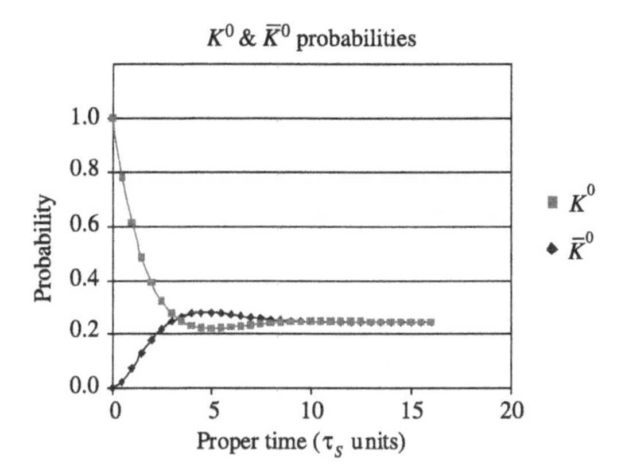

Fig. 12.1. Evolution of $\bar{K}^0$ and K^0 from an initially pure K^0 beam, using Eqs. (12.56) and (12.58) given in the text.

In this regime, the probability oscillates with a period $4\pi\tau_S$ with a damping factor $e^{-\frac{t}{2\tau_S}}$, and for $t > \tau_L \gg \tau_S$, the probability (12.5) takes the form

$$P(\bar{K}^0, t) \approx \frac{1}{4} e^{-\frac{t}{\tau_L}}, \tag{12.8}$$

so that the probability vanishes exponentially with the characteristic scale τ_L. There must therefore be a maximum in the $\bar{K}^0$ component. A plot of the full probability from Eq. (12.1) as a function of time is displayed in Fig. 12.1, and explicitly shows all these features. Note that the maximum in $\bar{K}^0$ occurs at $t \approx 4\tau_S$.

Problem 12.2 *Using the parameters η_{+-} and ϕ_{+-} of Eq.* (12.27), *derive an expression for the rate of $K^0 \to \pi^+\pi^-$ decay as a function of time. Assume that you start with a pure K^0 beam that develops according to Eq.* (12.55). *You may ignore the overall normalization of the decay rate.*

As discussed in the text, the states $|K_S^0\rangle$ and $|K_L^0\rangle$ in Eqs. (12.19) and (12.20) of the text coincide with those in Eq. (12.50), with the identification

$$p = 1 + \epsilon, \quad q = -1 + \epsilon. \tag{12.9}$$

Furthermore, from the parametrization in Eqs. (12.24) and (12.27) of the text, we can write

$$\epsilon = |\eta_{+-}|\ e^{i\phi_{+-}}. \tag{12.10}$$

With these identifications, the time evolution of the state $|K^0\rangle$ from Eq. (12.55) of the text (using the convention in the text of $c = 1$ in this discussion) becomes:

$$\begin{aligned}
|K^0(t)\rangle &= \frac{\sqrt{2(1+|\epsilon|^2)}}{2(1+\epsilon)}[e^{-\frac{i}{\hbar}(m_S - \frac{i}{2}\gamma_S)t}\ |K_S^0\rangle \\
&\quad + e^{-\frac{i}{\hbar}(m_L - \frac{i}{2}\gamma_L)t}\ |K_L^0\rangle] \\
&= \frac{\sqrt{2(1+|\eta_{+-}|^2)}}{2(1+|\eta_{+-}|e^{i\phi_{+-}})}[e^{-\frac{i}{\hbar}(m_S - \frac{i}{2}\gamma_S)t}\ |K_S^0\rangle \\
&\quad + e^{-\frac{i}{\hbar}(m_L - \frac{i}{2}\gamma_L)t}\ |K_L^0\rangle].
\end{aligned} \tag{12.11}$$

It follows now from Eq. (12.23) of the text that

$$\begin{aligned}
\langle\pi^+\pi^-|K^0(t)\rangle &= \frac{\sqrt{2(1+|\eta_{+-}|^2)}}{2(1+|\eta_{+-}|e^{i\phi_{+-}})}[e^{-\frac{i}{\hbar}(m_S - \frac{i}{2}\gamma_S)t}\ \langle\pi^+\pi^-|K_S^0\rangle \\
&\quad + e^{-\frac{i}{\hbar}(m_L - \frac{i}{2}\gamma_L)t}\ \langle\pi^+\pi^-|K_L^0\rangle] \\
&= \frac{\sqrt{2(1+|\eta_{+-}|^2)}}{2(1+|\eta_{+-}|e^{i\phi_{+-}})}\ \langle\pi^+\pi^-|K_S^0\rangle[e^{-\frac{i}{\hbar}(m_S - \frac{i}{2}\gamma_S)t} \\
&\quad + |\eta_{+-}|e^{-\frac{i}{\hbar}(m_L - \frac{i}{2}\gamma_L)t + i\phi_{+-}}].
\end{aligned} \tag{12.12}$$

This represents the time evolution for the transition amplitude for the decay

$$K^0 \to \pi^+ + \pi^-. \tag{12.13}$$

The time evolution for the rate of decay is obtained from

$$
\begin{aligned}
P(\pi^+ + \pi^-, t) &= |\langle \pi^+\pi^- | K^0(t)\rangle|^2 \\
&= \frac{2(1+|\eta_{+-}|^2)}{4|1+|\eta_{+-}|e^{i\phi_{+-}}|}\ |\langle \pi^+\pi^- | K_S^0\rangle|^2 \\
&\quad \times [e^{-\frac{1}{\hbar}\gamma_S t} + |\eta_{+-}|^2 e^{-\frac{1}{\hbar}\gamma_L t} + |\eta_{+-}| e^{-\frac{1}{2\hbar}(\gamma_S+\gamma_L)t} \\
&\quad \times (e^{\frac{i}{\hbar}(m_L-m_S)-i\phi_{+-}} + e^{-\frac{i}{\hbar}(m_L-m_S)+i\phi_{+-}})] \\
&= \frac{2(1+|\eta_{+-}|^2)}{4(1+|\eta_{+-}|^2+2|\eta_{+-}|\cos\phi_{+-})}\ |\langle \pi^+\pi^- | K_S^0\rangle|^2 \\
&\quad \times \left[e^{-\frac{t}{\tau_S}} + |\eta_{+-}|^2 e^{-\frac{t}{\tau_L}} + 2|\eta_{+-}| e^{-\frac{1}{2}\left(\frac{1}{\tau_S}+\frac{1}{\tau_L}\right)t} \right. \\
&\quad \left. \times \cos\left(\frac{\Delta m}{\hbar}t - \phi_{+-}\right)\right].
\end{aligned}
\tag{12.14}
$$

13. Standard Model I

Problem 13.1 *Prove that Eq.* (13.49) *follows from Eq.* (13.48).

From Eq. (13.47) of the text we note that with the choice of the minimum of the potential at

$$x_{\text{min}} = \sqrt{\frac{m\omega^2}{\lambda}}, \tag{13.1}$$

we have from Eq. (13.48) of the text

$$\begin{aligned}
V(x_{\text{min}} + x, y) &= -\frac{1}{2}m\omega^2((x_{\text{min}} + x)^2 + y^2) + \frac{\lambda}{4}((x_{\text{min}} + x)^2 + y^2)^2 \\
&= -\frac{1}{2}(x_{\text{min}}^2 + 2x_{\text{min}}x + x^2 + y^2) \\
&\quad + \frac{\lambda}{4}(x_{\text{min}}^2 + 2x_{\text{min}}x + x^2 + y^2)^2 \\
&= -\frac{1}{2}(x_{\text{min}}^2 + 2x_{\text{min}}x + x^2 + y^2) \\
&\quad + \frac{\lambda}{4}(x_{\text{min}}^4 + 4x_{\text{min}}^2x^2 + (x^2 + y^2)^2 + 2x_{\text{min}}^2(x^2 + y^2) \\
&\quad + 4x_{\text{min}}^3x + 4x_{\text{min}}x(x^2 + y^2)) \\
&= \left(-\frac{1}{2}m\omega^2x_{\text{min}}^2 + \frac{\lambda}{4}x_{\text{min}}^4\right) + (-m\omega^2 + \lambda x_{\text{min}}^2)x_{\text{min}}x \\
&\quad + \left(-\frac{1}{2}m\omega^2 + \lambda x_{\text{min}}^2 + \frac{\lambda}{2}x_{\text{min}}^2\right)x^2 \\
&\quad + \left(-\frac{1}{2}m\omega^2 + \frac{\lambda}{2}x_{\text{min}}^2\right)y^2 \\
&\quad + \lambda x_{\text{min}}x(x^2 + y^2) + \frac{\lambda}{4}(x^2 + y^2)^2
\end{aligned}$$

$$
\begin{aligned}
&= \left(-\frac{1}{2}m\omega^2 \times \frac{m\omega^2}{\lambda} + \lambda\left(\frac{m\omega^2}{\lambda}\right)^2\right) \\
&\quad + \left(-m\omega^2 + \lambda \times \frac{m\omega^2}{\lambda}\right) x_{\min} x \\
&\quad + \left(-\frac{1}{2}m\omega^2 + \frac{3\lambda}{2} \times \frac{m\omega^2}{\lambda}\right) x^2 \\
&\quad + \left(-\frac{1}{2}m\omega^2 + \frac{\lambda}{2} \times \frac{m\omega^2}{\lambda}\right) y^2 \\
&\quad + \lambda \times \sqrt{\frac{m\omega^2}{\lambda}} x(x^2+y^2) + \frac{\lambda}{4}(x^2+y^2)^2 \\
&= -\frac{m^2\omega^4}{4\lambda} + m\omega^2 x^2 + \sqrt{\lambda m\omega^2} x(x^2+y^2) \\
&\quad + \frac{\lambda}{4}(x^2+y^2)^2,
\end{aligned} \tag{13.2}
$$

which is the result in Eq. (13.49) of the text.

Problem 13.2 *According to the quark model, wave functions of baryons are antisymmetric in color. Construct a wave function for the Δ^{++} that is explicitly antisymmetric under the exchange of any two of its quark constituents in color space.*

As discussed in Sec. 13.5 of the text, the description of Δ^{++} in terms of the quark model naturally leads to the need for a new quantum number "color". From Sec. 9.6 as well as from Table 10.2 of the text, we see that Δ^{++} is a nonstrange baryon with quantum numbers

$$
B = 1, \quad Q = 2, \quad I = \frac{3}{2}, \quad I_3 = \frac{3}{2}, \quad J = \frac{3}{2}, \quad J_z = \frac{3}{2}, \quad S = 0. \tag{13.3}
$$

In terms of the quark model, we can therefore describe this state by the three-quark system:

$$
|\Delta^{++}\rangle = |uuu\rangle, \tag{13.4}
$$

where the spins of the u quarks are all parallel and point along the same direction. This state satisfies the quantum numbers given in (13.3). However, in the ground state, where $\ell = 0$, the wave function for such a state will be symmetric in both spin and isospin spaces (space part is symmetric since $(-1)^\ell = 1$ for $\ell = 0$). On the other hand, the Pauli principle requires that a composite state of three identical fermions be antisymmetric under exchange of any two fermions. This suggests that quarks carry an additional quantum number, and that the wave function for Δ^{++} is antisymmetric in this "color" quantum number, thereby making the total wave function antisymmetric.

Letting

$$|u^a\rangle, \quad a = r, b, g, \tag{13.5}$$

denote the three color states of the u quark, a normalized state manifestly antisymmetric under the exchange of any two color indices can then be constructed as

$$\begin{aligned}|\Delta^{++}\rangle = \frac{1}{\sqrt{6}}(&|u^r u^b u^g\rangle - |u^r u^g u^b\rangle + |u^b u^g u^r\rangle - |u^b u^r u^g\rangle \\ &+ |u^g u^r u^b\rangle - |u^g u^b u^r\rangle).\end{aligned} \tag{13.6}$$

Despite the fact that their constituents carry color quantum numbers, baryons (e.g. Δ^{++}) do not. In fact, all observed hadrons are color-neutral or "singlets" in color. This can be attributed to the "saturated" color structure of the quark states, wherein all three colors contribute to each part of the wave function.

14. Standard Model and Confrontation with Data

Problem 14.1 *The mass of the top quark is larger than that of the W boson. It is consequently not surprising that the top quark decays into a W and a b quark ($t \to W + b$). The expected width (Γ) of the top quark in the Standard Model is $\approx 1.5\,\text{GeV}$.* (a) *What can you say about the lifetime of the top quark?* (b) *If QCD color interactions can be characterized by the fly-by time of two hadrons (time needed to exchange gluons), what is the ratio of lifetime to interaction time for top quarks?* (c) *Because of the rapid fall-off in parton distributions $f(x, \mu)$ with increasing x, the peak of the production cross section for $t\bar{t}$ events in $p\bar{p}$ collisions occurs essentially at threshold. What is the typical momentum of the b quark in $p\bar{p}$ collisions that yield $t\bar{t}$ events.* (d) *What are the typical x values of the colliding partons that can produce $t\bar{t}$ events at the Tevatron ($\sqrt{s} = 2\,\text{TeV}$)? What about at the LHC ($\sqrt{s} = 14\,\text{TeV}$)? (Hint: $\hat{s} = x_a x_b s$, where $\hat{s}$ is the value of the square of the energy in the rest frame of the partonic collision of a and b. Can you prove this?)*

(a) In discussions about resonances in Sec. 9.6 of the text, we saw in Eqs. (9.27) and (9.33) that for a state with decay width Γ, the mean life is given by (we are restoring factors of c that were set to unity in Chap. 14)

$$\tau = \frac{\hbar}{\Gamma c}. \tag{14.1}$$

For the top quark, we are given that

$$\Gamma_{\text{top}} \approx 1.5\,\text{GeV}/c^2. \tag{14.2}$$

It therefore follows that the mean life of the top quark has the value

$$\tau_{\text{top}} = \frac{\hbar}{\Gamma_{\text{top}} c} = \frac{\hbar c}{\Gamma_{\text{top}} c^2} \times \frac{1}{c}$$
$$\approx \frac{200\,\text{MeV} - \text{F}}{1.5\,\text{GeV}} \times \frac{1}{3 \times 10^{10}\,\text{cm/sec}}$$
$$\approx \frac{2 \times 10^2 \times 10^{-13}\,\text{MeV-cm}}{4.5 \times 10^{13}\,\text{MeV-cm/sec}}$$
$$\approx 4.4 \times 10^{-26}\,\text{sec}. \tag{14.3}$$

(b) The typical size of hadrons is about the size of a nucleon, that is,

$$R \approx 1.0 \times 10^{-13}\,\text{cm}. \tag{14.4}$$

Consequently, the fly-by time between two hadrons is given by

$$\tau_{\text{fly-by}} \approx \frac{R}{c} = \frac{1.0 \times 10^{-13}\ \text{cm}}{3 \times 10^{10}\,\text{cm/sec}} \approx 3 \times 10^{-24}\ \text{sec}. \tag{14.5}$$

This leads to the ratio

$$\frac{\tau_{\text{top}}}{\tau_{\text{fly-by}}} \approx \frac{4.4 \times 10^{-26}\ \text{sec}}{3 \times 10^{-24}\,\text{sec}} \approx 1.5 \times 10^{-2}. \tag{14.6}$$

This suggests that a top quark decays before it has time to interact.

(c) When $t\bar{t}$ pairs are produced in collisions:

$$p + \bar{p} \to t + \bar{t}, \tag{14.7}$$

they subsequently decay through the channels

$$t \to W^+ + b, \quad \bar{t} \to W^- + \bar{b}. \tag{14.8}$$

If the $t\bar{t}$ pair is produced at threshold, then the t and $\bar{t}$ quarks are essentially at rest. Consequently, the Q value for the reaction (see, for example, Eq. (4.4) of the text) is

$$Q = T_W + T_b = (M_t - M_W - M_b)c^2. \tag{14.9}$$

From Table 9.5 and Sec. 13.11 of the text, we have

$$M_t \approx 175\,\text{GeV}/c^2, \quad M_W \approx 80.4\ \text{GeV}/c^2, \quad M_b \approx 4.2\,\text{GeV}/c^2. \tag{14.10}$$

Therefore,

$$Q = (M_t - M_W - M_b)c^2 \approx (175 - 80.4 - 4.2)\,\mathrm{GeV}/c^2 \times c^2$$
$$= 90.4\,\mathrm{GeV}. \tag{14.11}$$

This is a large amount of energy, but we see from the masses of the particles that we can treat the W bosons as almost nonrelativistic, but the b quarks as (ultra) relativistic objects. The situation is therefore quite similar to that of Problem 4.3.

Since the top quark decays at rest, momentum conservation leads to

$$\vec{p}_W = -\vec{p}_b. \tag{14.12}$$

Substituting this into Eq. (14.9) we obtain

$$T_W + T_b = Q$$

$$\text{or} \quad \frac{p_W^2}{2M_W} + T_b = Q$$

$$\text{or} \quad p_b^2c^2 + 2T_bM_Wc^2 - 2QM_Wc^2 = 0$$

$$\text{or} \quad T_b^2 + 2T_bM_bc^2 + 2T_bM_Wc^2 - 2QM_Wc^2 = 0$$

$$\text{or} \quad T_b^2 + 2T_b(M_W + M_b)c^2 - 2QM_Wc^2 = 0$$

$$\text{or} \quad T_b = \frac{-2(M_W + M_b)c^2 \pm \sqrt{4(M_W + M_b)^2c^4 + 8QM_Wc^2}}{2}$$

$$= -(M_W + M_b)c^2 \pm (M_W + M_b)c^2\left(1 + \frac{2QM_Wc^2}{(M_W + M_b)^2c^4}\right)^{\frac{1}{2}}, \tag{14.13}$$

where we have used Eq. (A.10) of the text in the intermediate steps. Clearly, only one solution is physical (positive T_b):

$$T_b = (M_W + M_b)c^2\left[\left(1 + \frac{2QM_Wc^2}{(M_W + M_b)^2c^4}\right)^{\frac{1}{2}} - 1\right]$$

$$\approx 84.6\,\mathrm{GeV}\left[\left(1 + \frac{2 \times 90.4\ \mathrm{GeV} \times 80.4\,\mathrm{GeV}}{(84.6\ \mathrm{GeV})^2}\right)^{\frac{1}{2}} - 1\right]$$

$$\approx 84.6\,\text{GeV}\left[(1+2.03)^{\frac{1}{2}}-1\right] = 84.6\,\text{GeV} \times 0.74$$
$$\approx 62.6\,\text{GeV}. \qquad (14.14)$$

We see that the b quark carries away most of the released energy in the process, and its momentum follows from the definition in Eq. (A.10) of the text:

$$p_b = \frac{1}{c}\sqrt{(T_b + M_b c^2)^2 - (M_b c^2)^2}$$
$$\approx \frac{1}{c}\sqrt{(62.6+4.2)^2\,\text{GeV}^2 - (4.2)^2\,\text{GeV}^2} \approx 66.7\,\text{GeV}/c. \qquad (14.15)$$

(d) Let us consider the process

$$p + \bar{p} \rightarrow t + \bar{t}, \qquad (14.16)$$

in the center of mass. If the $t\bar{t}$ pair is produced at rest, then it is clear [see Eq. (1.65) of the text] that the total energy in the center of mass is:

$$E_{\text{CM}}^{\text{TOT}} = 2M_t c^2 = 350\,\text{GeV} = 0.35\,\text{TeV}. \qquad (14.17)$$

To simplify the problem, let us assume that the process takes place through the interaction of a pair of almost massless partons (a parton and an antiparton), so that we can write the reaction as

$$\text{“}u\text{”} + \text{“}\bar{u}\text{”} \rightarrow t + \bar{t}. \qquad (14.18)$$

In the center of mass, the collision will appear kinematically symmetric, and we can define

$$P^{\mu}_{\text{“}u\text{”}} = xP^{\mu}_{\text{CM}}, \quad P^{\mu}_{\text{“}\bar{u}\text{”}} = x\bar{P}^{\mu}_{\text{CM}}, \qquad (14.19)$$

where x denotes the fraction of the four-momentum of the proton that is carried by the parton. In this rest frame, the spatial momenta of the two partons will be equal and opposite, and the available energy in the collision will be given by

$$(P_{\text{"}u\text{"}} + P_{\text{"}\bar{u}\text{"}})^2 = (xE_p^{\rm CM} + xE_{\bar{p}}^{\rm CM})^2 = 4x^2 s, \tag{14.20}$$

where s represents the square of the available beam energy. To produce a $t\bar{t}$ pair, we must have

$$\begin{aligned} 4x^2 s &= (0.35\,\text{TeV})^2 \\ \text{or}\quad x &= \frac{0.35\,\text{TeV}}{2\sqrt{s}}. \end{aligned} \tag{14.21}$$

At Fermilab, we have $\sqrt{s} = 2\,\text{TeV}$, so that

$$x_{\text{Fermilab}} = \frac{0.35\,\text{TeV}}{2 \times 2\,\text{TeV}} \approx 0.09, \tag{14.22}$$

while at the LHC, with $\sqrt{s} = 14\,\text{TeV}$, this leads to

$$x_{\text{LHC}} = \frac{0.35\,\text{TeV}}{2 \times 14\,\text{TeV}} \approx 0.012. \tag{14.23}$$

Problem 14.2 *In discussing weak decays proceeding through W or Z bosons, we have focused primarily on the fundamental transitions among quarks and leptons. However, such decays often involve hadrons that contain spectator quarks, in addition to the partons that participate in the weak interaction (see Fig. 14.3). For example, Fig. 14.5 shows a diagram for the decay of a K^0 into a $\pi^+\pi^-$ pair. Using similar quark-line diagrams, draw processes for the following decays:* (a) $K^+ \to \pi^+ + \pi^0$, (b) $n \to p + e^- + \bar{\nu}_e$, (c) $\pi^+ \to \mu^+ + \nu_\mu$, (d) $K^0 \to \pi^- + e^+ + \nu_e$.

See Figs. 14.1–14.4 that provide the diagrams for decays 14.2(a)–(d), respectively.

Problem 14.3 *Draw quark-line diagrams for the following reactions:* (a) $\pi^- + p \to \Lambda^0 + K^0$, (b) $\pi^+ + p \to \Sigma^+ + K^+$, (c) $\pi^+ + n \to \pi^0 + p$, (d) $p + p \to \Lambda^0 + K^+ + p$, (e) $\bar{p} + p \to K^+ + K^-$.

See Figs. 14.5–14.9 that provide the diagrams for decays 14.3(a)–(e), respectively.

Problem 14.4 *Draw quark-line diagrams for the following weak interactions, and include any required intermediate W or Z bosons:* (a) $\nu_e + n \to \nu_e + n$, (b) $\bar{\nu}_\mu + p \to \mu^+ + n$, (c) $\pi^- + p \to \Lambda^0 + \pi^0$.

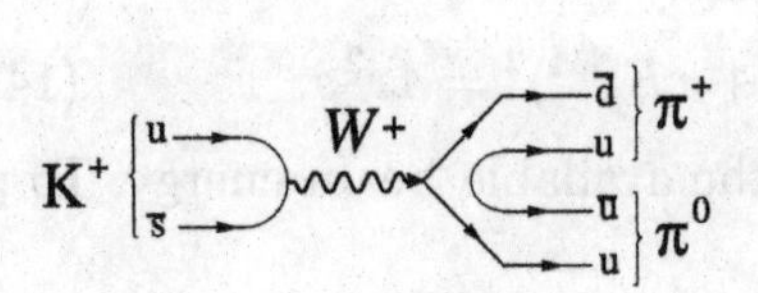

Fig. 14.1. Decay 14.2a.

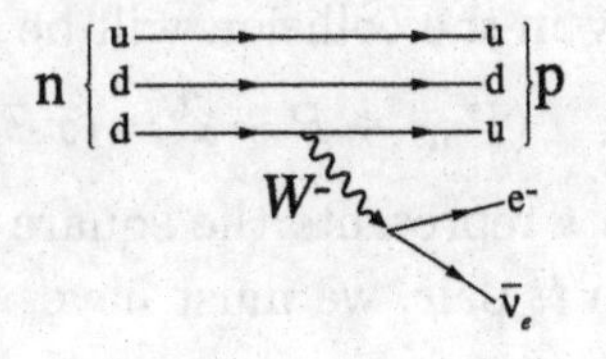

Fig. 14.2. Decay 14.2b.

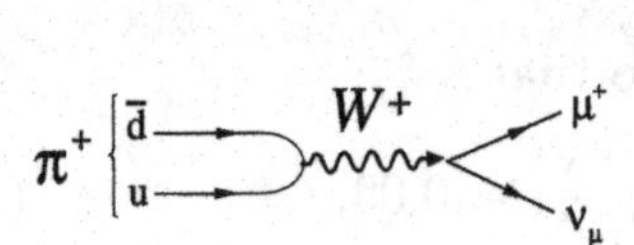

Fig. 14.3. Decay 14.2c.

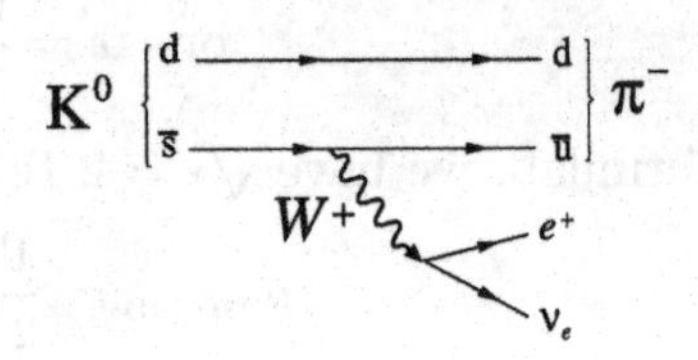

Fig. 14.4. Decay 14.2d.

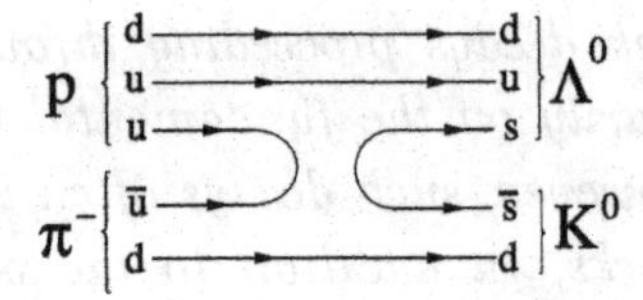

Fig. 14.5. Reaction 14.3a.

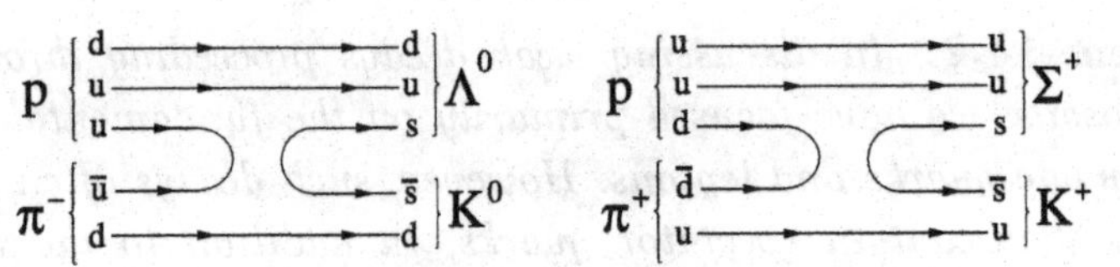

Fig. 14.6. Reaction 14.3b.

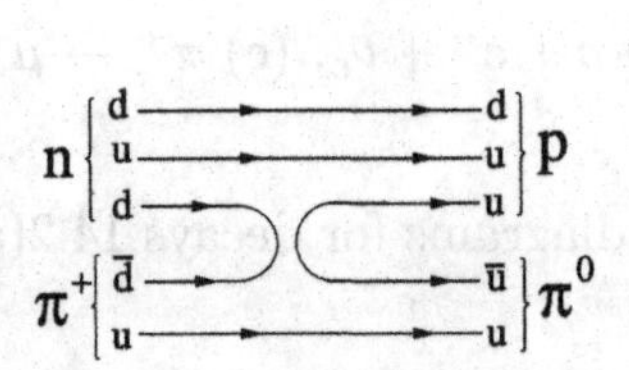

Fig. 14.7. Reaction 14.3c.

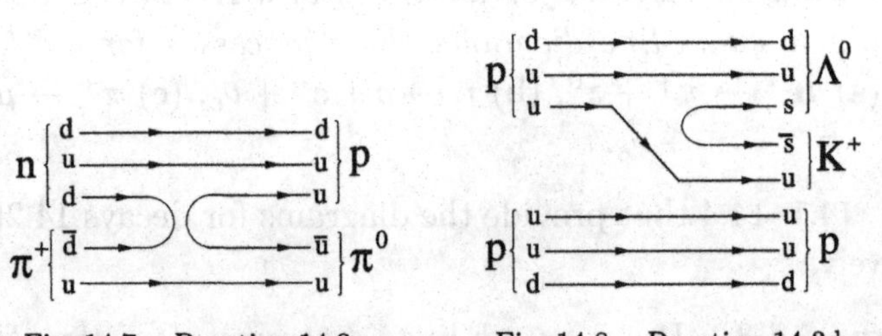

Fig. 14.8. Reaction 14.3d.

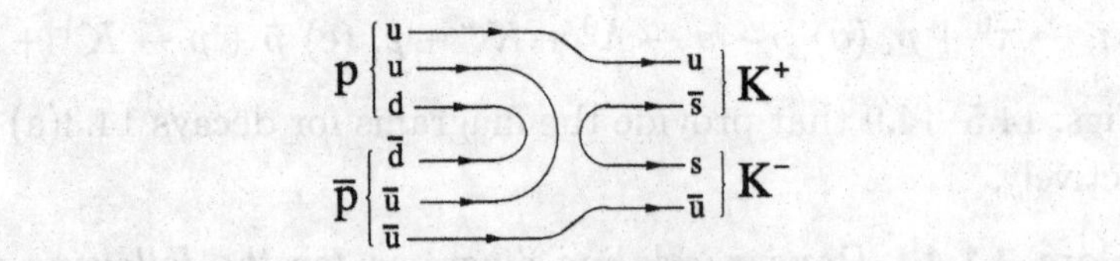

Fig. 14.9. Reaction 14.3e.

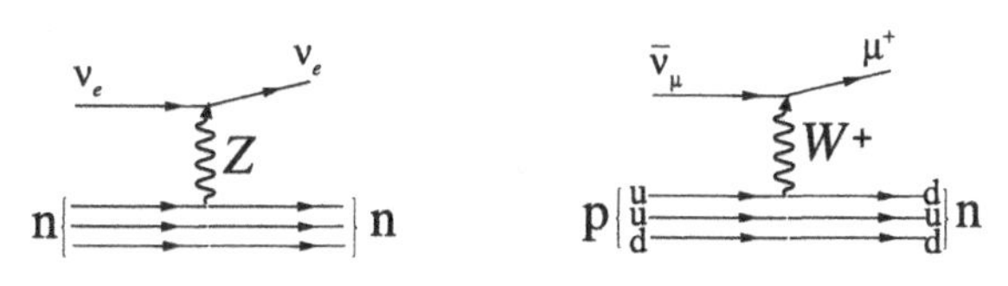

Fig. 14.10. Reaction 14.4a. Fig. 14.11. Reaction 14.4b.

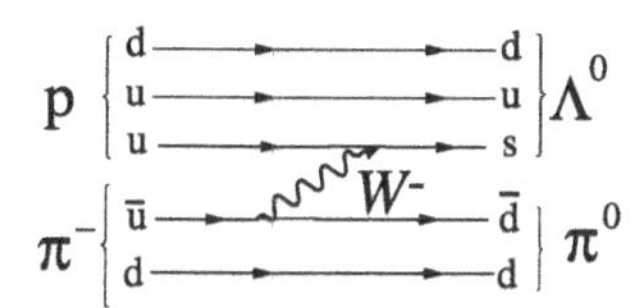

Fig. 14.12. Reaction 14.4c.

See Figs. 14.10–14.12 that provide the diagrams for decays 14.4(a)–(c), respectively.

Problem 14.5 *One of the main reasons for the introduction of the GIM mechanism was the need to suppress flavor-changing neutral currents, in order to reduce the rate for* $K_L^0 \to \mu^+\mu^-$ *to its observed small value.* (a) *Draw the quark-line diagram for this transition involving W bosons (via a higher-order box diagram), and the possible contribution from* Z^0 *exchange.* (b) *Show that the* Z^0 *contribution vanishes once the weak states of Eq.* (14.8) *are used to calculate that contribution. (Hint: Contrast the transition elements* $\langle d'\bar{d}'|Z^0\rangle$, $\langle s'\bar{s}'|Z^0\rangle$, *and* $\langle d'\bar{s}'|Z^0\rangle$ *by considering* $\langle d'\bar{d}'\rangle$, $\langle s'\bar{s}'\rangle$, *and* $\langle d'\bar{s}'\rangle$.)

The reaction $K^0 \to \mu^+\mu^-$ can take place through the second-order weak transitions described in the two box diagrams shown in Figs. 4.13–4.14. The diagrams differ only in the nature of the exchanged quarks. In addition to a u-quark or a c-quark, a t-quark can also be exchanged. These weak transitions are greatly suppressed. Figure 4.15 shows a diagram that is a first-order (leading) weak transition via an "s-channel" Z boson. This strangeness-changing neutral-current diagram is ostensibly allowed, and should therefore provide a large rate that can be measured for this process.

However, this process is suppressed by the "GIM" mechanism of mixing of weak eigenstates via the unitary CKM matrix. The important parameter is the Cabbibo angle θ_C.

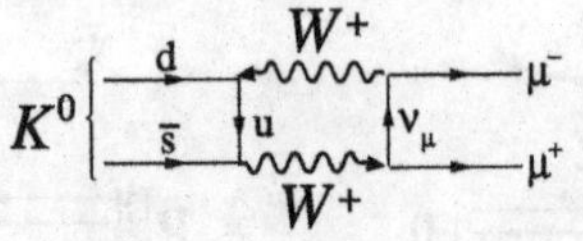

K^0 d $\bar{s}$ u W^+ W^+ ν_μ μ^- μ^+

Fig. 14.13. Box diagram decay with u-quark exchange.

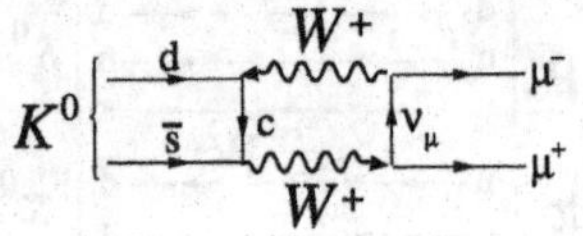

K^0 d $\bar{s}$ c W^+ W^+ ν_μ μ^- μ^+

Fig. 14.14. Box diagram decay with c quark exchange.

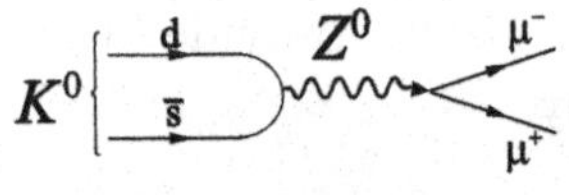

K^0 d $\bar{s}$ Z^0 μ^- μ^+

Fig. 14.15. Decay via s-channel Z boson contribution.

The suppression of strangeness-changing neutral currents via the GIM mechanism comes about because the weak eigenstates (d', s') correspond to a specific mixing of the strong eigenstates (d, s), and vice versa:

$$\begin{pmatrix} d \\ s \end{pmatrix} = \begin{pmatrix} \cos\theta_c & -\sin\theta_c \\ \sin\theta_c & \cos\theta_c \end{pmatrix} \begin{pmatrix} d' \\ s' \end{pmatrix}$$

with d and s quark states comprised of the following mixtures of the weak eigenstates

$$d = d' \cos\theta_c - s' \sin\theta_c,$$
$$\bar{s} = \bar{d}' \sin\theta_c + \bar{s}' \cos\theta_c.$$

Using these transformations, the coupling of the Z boson to the strong interaction state $K^0 = (d, \bar{s})$ can be expressed in terms of the eigenstates of the weak interaction as:

$$\begin{aligned} \langle d, \bar{s}|Z\rangle = \langle d', \bar{d}'|Z\rangle \cos\theta_c \sin\theta_c - \langle s', \bar{s}'|Z\rangle \cos\theta_c \sin\theta_c \\ + \langle d', \bar{s}'|Z\rangle \cos^2\theta_c - \langle s', \bar{d}'|Z\rangle \sin^2\theta_c. \end{aligned}$$

Because of the orthogonality of the weak eigenstates, the amplitudes for transitions between states of different flavor $\langle d', \bar{s}'|Z\rangle$ and

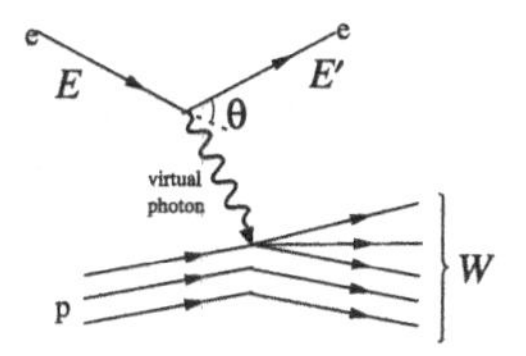

Fig. 14.16. Deep inelastic electron scattering.

$\langle s', \bar{d}'|Z\rangle$ vanish. The probabilities are equal for the two amplitudes $\langle d', \bar{d}'|Z\rangle = \langle s', \bar{s}'|Z\rangle = 1/\sqrt{2}$, and therefore

$$\langle d, \bar{s}|Z\rangle = \frac{1}{\sqrt{2}}(\langle d', \bar{d}'|Z\rangle \cos\theta_c \sin\theta_c - \langle s', \bar{s}' Z\rangle \cos\theta_c \sin\theta_c) = 0.$$

Thus the Z contribution to the rate for the decay is also expected to be very highly suppressed.

Problem 14.6 *Consider the scattering of an electron from a proton (of mass m_p), as shown in Fig.* 14.3. *Let W be the invariant mass of the entire recoiling hadronic system, and Q and P (except for multiplicative factors of c) the four-momenta of the exchanged vector boson and target proton, respectively, and E, E' and θ the incident energy, the scattered energy and scattering angle of the electron in the laboratory (i.e. the rest frame of the proton). Defining Q^2 as $(\vec{k}' - \vec{k})^2 c^2 - \nu^2$, where $\vec{k}'$ and $\vec{k}$ are the three-momenta of the scattered and incident electron, and ν is the difference in electron energy, show that for very high energies,* (a) $Q^2 = 4EE' \sin^2\frac{\theta}{2}$, (b) $W^2 = m_p^2 + \frac{2m_p\nu}{c^2} - \frac{Q^2}{c^4}$. (c) *What is the smallest value that W can assume? What type of scattering does that correspond to?* (d) *What is the largest Q^2 possible? What does that correspond to? What is the mass of the vector boson in this case?* (e) *What is the largest possible value of W?*

We are studying the process

$$e^- + p \to e^- + X, \tag{14.24}$$

where X denotes the entire hadronic system of mass W produced in the scattering (see Fig. 14.16).

Let us denote by k^μ, P^μ, k'^μ and P_X^μ the four-momenta of the incident electron, the target proton, the scattered electron and the produced hadronic system, respectively. Since the proton is initially at rest, we can represent the four-vectors as

$$\begin{aligned} k^\mu &= \left(\frac{E}{c}, \vec{k}\right), \qquad P^\mu = (m_p c, 0), \\ k'^\mu &= \left(\frac{E'}{c}, \vec{k}'\right), \quad P_X^\mu = \left(\frac{E_X}{c}, \vec{P}_X\right). \end{aligned} \tag{14.25}$$

Conservation of energy and momentum leads to

$$k^\mu + P^\mu = k'^\mu + P_X^\mu, \tag{14.26}$$

which can be written as

$$\begin{aligned} E + m_p c^2 &= E' + E_X, \\ \vec{k} &= \vec{k}' + \vec{P}_X. \end{aligned} \tag{14.27}$$

The scattering angle θ corresponds to the angle between the vectors $\vec{k}$ and $\vec{k}'$.

(a) The momentum transfer in the process is given by (see Eq. (1.67) of the text and the discussion there)

$$q^\mu = (k^\mu - k'^\mu) = \left[\frac{\nu}{c}, (\vec{k} - \vec{k}')\right] = \left[\frac{(E - E')}{c}, (\vec{k} - \vec{k}')\right], \tag{14.28}$$

which leads to the invariant

$$\begin{aligned} Q^2 &= -c^2 q^2 = -(E - E')^2 + (\vec{k} - \vec{k}')^2 c^2 \\ &= \vec{k}^2 c^2 + \vec{k}'^2 c^2 - 2\vec{k} \cdot \vec{k}' c^2 - (E^2 + E'^2 - 2EE') \\ &= -(E^2 - \vec{k}^2 c^2) - (E'^2 - \vec{k}'^2 c^2) + 2EE' - 2|\vec{k}||\vec{k}'|c^2 \cos\theta \\ &= -m_e^2 c^4 - m_e^2 c^4 + 2EE' - 2|\vec{k}|\vec{k}'|c^2 \cos\theta \\ &= -2m_e^2 c^4 + 2EE' - 2|\vec{k}||\vec{k}'|c^2 \cos\theta. \end{aligned} \tag{14.29}$$

This is a general result. However, if the electrons are extremely energetic, we can treat them as ultrarelativistic and neglect their masses, in which case we can approximate

$$|\vec{k}|c \approx E, \quad |\vec{k}'| \approx E', \tag{14.30}$$

and get

$$Q^2 = 2EE' - 2EE'\cos\theta = 2EE'(1-\cos\theta) = 4EE'\sin^2\frac{\theta}{2}, \tag{14.31}$$

which is the desired result.

(b) From Eq. (14.26), we have that

$$P_X^\mu = P^\mu + k^\mu - k'^\mu. \tag{14.32}$$

From the definition of invariant mass in Eq. (1.64) of the text (where it is denoted as $\frac{\sqrt{s}}{c^2}$, we obtain for the square of the invariant mass of system X:

$$\begin{aligned}
W^2 &= \frac{1}{c^2}\,P_X^2 = \frac{1}{c^2}\left(P + (k - k')\right)^2 \\
&= \frac{1}{c^2}\left[\left(m_p c + \frac{(E - E')}{c}\right)^2 - (\vec{k} - \vec{k}')^2\right] \\
&= \frac{1}{c^2}\left[m_p^2c^2 + 2m_p(E - E') + \frac{1}{c^2}\left((E - E')^2 - (\vec{k} - \vec{k}')^2c^2\right)\right] \\
&= m_p^2 + \frac{2m_p\nu}{c^2} - \frac{Q^2}{c^4},
\end{aligned} \tag{14.33}$$

where $\nu = E - E'$. This is the desired result.

(c) From the first Eq. (14.27), we obtain:

$$\begin{aligned}
\vec{P}_X &= \vec{k} - \vec{k}' \\
\text{or} \quad \vec{P}_X^2 &= \vec{k}^2 + \vec{k}'^2 - 2\vec{k}\cdot\vec{k}' \\
&\approx \frac{1}{c^2}(E^2 + E'^2 - 2EE'\cos\theta),
\end{aligned} \tag{14.34}$$

where we have used the fact that the electron is ultrarelativistic. Using this in the first relationship of Eq. (14.27) leads to

$$
\begin{aligned}
& E + m_p c^2 = E' + E_X = E' + \sqrt{m_X^2 c^4 + \vec{P}_X^2 c^2} \\
\text{or} \quad & (E - E' + m_p c^2)^2 = m_X^2 c^4 + \vec{P}_X^2 c^2 \\
\text{or} \quad & E^2 + E'^2 + m_p^2 c^4 + 2E m_p c^2 - 2E'(E + m_p c^2) \\
& \quad = m_X^2 c^4 + E^2 + E'^2 - 2EE' \cos\theta \\
\text{or} \quad & 2E m_p c^2 + (m_p^2 - m_X^2) c^4 = 2E'(E(1 - \cos\theta) + m_p c^2) \\
\text{and} \quad & E' = \frac{2E m_p c^2 + (m_p^2 - m_X^2) c^4}{2(E(1 - \cos\theta) + m_p c^2)}. \qquad (14.35)
\end{aligned}
$$

Thus, the value of E' is a function of the scattering angle θ. By definition, $E' \geq 0$, and from the above equation we can see that E' has its maximum:

$$E'_{\text{max}} = E, \qquad (14.36)$$

which is reached when $\theta = 0$ and $m_X = m_p$. This corresponds to elastic scattering off a proton at rest. Using Eq. (14.31), we can write Eq. (14.33) as

$$
\begin{aligned}
W^2 &= m_p^2 + \frac{2m_p(E - E')}{c^2} - \frac{4EE'}{c^4} \sin^2\frac{\theta}{2} \\
&= m_p^2 + \frac{2m_p E}{c^2} - 2E'\left(\frac{m_p}{c^2} + \frac{2E}{c^4} \sin^2\frac{\theta}{2}\right). \qquad (14.37)
\end{aligned}
$$

For a fixed value of E, W^2 is a function of E' and θ. We can find the extremes of this function by taking derivatives and setting them to zero. But this can be done more easily recognizing that W^2 is a linear function of E' (with a negative sign), and the minimum value of W^2 occurs when $E' = E'_{\text{max}} = E$, which, as we have seen, happens for $\theta = 0$. For these values, we obtain

$$W^2_{\text{min}} = m_p^2 + \frac{2m_p E}{c^2} - 2E\left(\frac{m_p}{c^2}\right) = m_p^2. \qquad (14.38)$$

The minimum in invariant mass is therefore obtained when the electron scatters forward in an elastic manner and the target proton remains at rest.

(d) From Eqs. (14.31) and (14.35), we have that

$$Q^2 = 4EE' \sin^2 \frac{\theta}{2} = \frac{2E(2Em_pc^2 + (m_p^2 - m_X^2)c^4)}{(E(1-\cos\theta) + m_pc^2)}. \tag{14.39}$$

It follows that the maximum value of Q^2, obtained when $\theta = \pi, m_X = m_p$, has the value

$$Q^2_{\text{max}} = \frac{4E^2 m_p c^2}{2E + m_p c^2}, \tag{14.40}$$

This corresponds to the case where the electron is scattered backwards.

(e) From (14.37), we see that the maximum value for the invariant mass will be obtained for $E' = 0$, in which case we have

$$W^2_{\text{max}} = m_p^2 + \frac{2m_pE}{c^2}. \tag{14.41}$$

Here, after hitting the target, the electron comes to rest and the produced hadronic system takes away all the momentum.

Problem 14.7 *Now consider the scattering of Problem* 14.6 *in a frame in which the proton has an exceedingly large three-momentum, so that its mass m_p can be ignored, as can the transverse momenta of all its partons. Now, suppose that the collision involves a parton carrying a fraction x of the proton's four-momentum, and that it absorbs the exchanged "four-momentum" Q.* (a) *First, show that in the laboratory frame $Q \cdot P = m_p\nu c^2$.* (b) *Now, prove that, for very large Q^2 (corresponding to deep-inelastic scattering), and in particular when $Q^2 \gg x^2 m_p^2 c^4$, $x = \frac{Q^2}{2m_p\nu c^2}$.* (c) *Plot $\frac{Q^2}{c^4}$ as a function of $\frac{2m_p\nu}{c^2}$ for $W = m_p$, $W = \sqrt{5}m_p$ and $W = 3m_p$.* (d) *Indicate the regions in* (c) *that correspond to $x < 1$, $x < 0.5$ and $x < 0.1$.* (e) *Identify the approximate location of the point corresponding to $E = 10\,\text{GeV}$, $E' = 1\,\text{GeV}$ and $\theta = \frac{\pi}{3}$, and the point corresponding to $E = 10\,\text{GeV}$, $E' = 4\,\text{GeV}$ and $\theta = \frac{\pi}{6}$, on the plot in part* (c).

(a) From Eq. (14.25) of the previous problem, we have that in the frame in which the proton is at rest:

$$q^\mu = (k^\mu - k'^\mu) = \left(\frac{(E-E')}{c}, (\vec{k} - \vec{k}')\right),$$
$$P^\mu = (m_p c, 0), \tag{14.42}$$

(Q^μ in the problem should be identified with q^μ. The four-vectors used in this specific problem are defined with factors of c different from what is given in the Appendix.) Through direct evaluation, we have

$$P \cdot qc^2 = m_p c \times \frac{(E-E')}{c} \times c^2 = m_p \nu c^2. \tag{14.43}$$

This is a Lorentz invariant quantity, and is therefore independent of the frame of reference. In particular, it holds even in the frame in which the proton has an exceedingly large spatial momentum (the "infinite momentum frame").

(b) Let us assume that only one parton of the proton participates in the reaction, and that its four-momentum is given by (see Fig. 14.17):

$$P^\mu_{\text{parton}} = xP^\mu, \tag{14.44}$$

where x represents the fraction of the proton's four-momentum that is carried by the parton. The effective reaction (see accompanying Fig. 14.17) is given by

$$e^- + \text{parton} \to e^- + \text{parton}. \tag{14.45}$$

Denoting by $P^\mu_{\text{parton}}, P'^\mu_{\text{parton}}$ the four-momenta of the incident and the final-state partons, respectively, conservation of energy-momentum

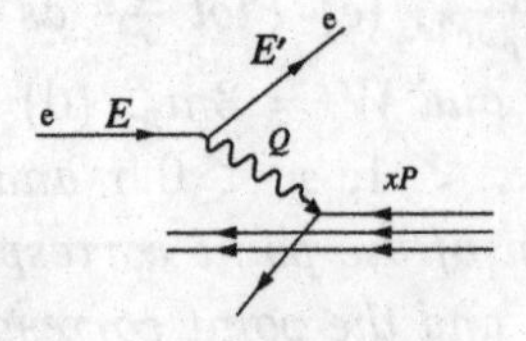

Fig. 14.17. Deep-inelastic scattering (DIS) in the infinite momentum frame.

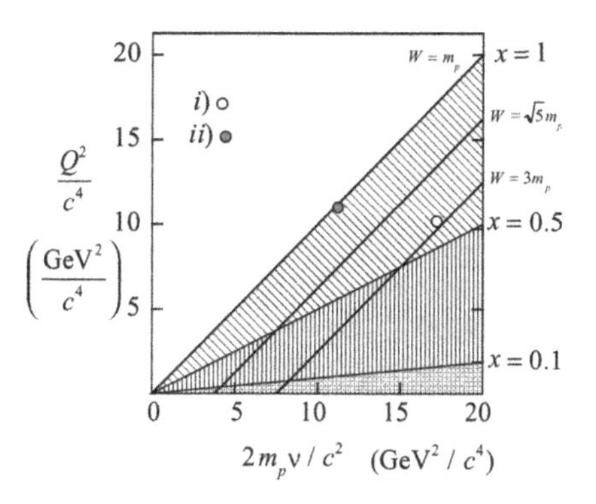

Fig. 14.18. Kinematics of deep-inelastic scattering.

leads to

$$
\begin{aligned}
P^{\mu}_{\text{parton}} + k^{\mu} &= P'^{\mu}_{\text{parton}} + k'^{\mu} \\
\text{or} \quad (xP^{\mu} + (k^{\mu} - k'^{\mu})) &= P'^{\mu}_{\text{parton}},
\end{aligned}
\tag{14.46}
$$

where we have used Eq. (14.44). If we assume the parton to be massless, then squaring the above relationship we obtain

$$
\begin{aligned}
&(xP + (k - k'))^2 = P'^2 \approx 0 \\
\text{or} \quad &(xP + q)^2 = x^2P^2 + q^2 + 2xP \cdot q \approx 0 \\
\text{and} \quad &x^2 m_p^2 c^2 - \frac{Q^2}{c^2} + 2xm_p\nu \approx 0,
\end{aligned}
\tag{14.47}
$$

where we have used Eq. (14.43) as well as the definition in (14.29). If we assume that

$$
Q^2 \gg x^2 m_p^2 c^4, \tag{14.48}
$$

then, we can neglect the first term in the above equation and this leads to the expected result

$$
x = \frac{Q^2}{2m_p\nu c^2}. \tag{14.49}
$$

(c) The plots of interest are given in Fig. 14.18.

15. Beyond the Standard Model

Problem 15.1 *Show that Eq.* (15.22) *follows from the definitions given in Eq.* (15.21).

The basic commutation relationships between the bosonic and fermionic operators are given in Eqs. (15.9) and (15.14) of the text. The substantive relationships take the forms

$$\left[a_B, a_B^\dagger\right] = 1, \quad \left[a_F, a_F^\dagger\right]_+ = a_F a_F^\dagger + a_F^\dagger a_F = 1, \tag{15.1}$$

and the fermionic properties are reflected in the relationships

$$[a_F, a_F]_+ = 0 = \left[a_F^\dagger, a_F^\dagger\right]_+ . \tag{15.2}$$

The supersymmetry charges are defined in Eq. (15.21) of the text, and have the forms

$$Q_F = a_B^\dagger a_F, \quad Q_F^\dagger = a_F^\dagger a_B. \tag{15.3}$$

It follows, therefore, that

$$\begin{aligned}
\left[Q_F, Q_F^\dagger\right]_+ &= Q_F Q_F^\dagger + Q_F^\dagger Q_F \\
&= a_B^\dagger a_F a_F^\dagger a_B + a_F^\dagger a_B a_B^\dagger a_F \\
&= a_B^\dagger \left(1 - a_F^\dagger a_F\right) a_B + a_F^\dagger \left(1 + a_B^\dagger a_B\right) a_F \\
&= a_B^\dagger a_B + a_F^\dagger a_F - a_B^\dagger a_F^\dagger a_F a_B + a_B^\dagger a_F^\dagger a_F a_B \\
&= a_B^\dagger a_B + a_F^\dagger a_F = \frac{1}{\hbar\omega}\, H,
\end{aligned} \tag{15.4}$$

where, in the intermediate steps, we have used the above relationships, as well as the fact that the bosonic and fermionic operators commute. This is the desired relationship in Eq. (15.22) of the text.

Problem 15.2 *Using dimensional analysis, and the known value of* $G_{\rm N}$*, show that you can write* $G_{\rm N} = \frac{\hbar c}{M_{\rm P}^2}$*, where* $M_{\rm P}$ *is the Planck mass or scale. What is the value of* $M_{\rm P}$ *in* GeV *units? Applying the uncertainty principle, you can define a Planck length and a Planck time, as in Eq.* (15.25)*. What are these values in* cm *and* sec, *respectively?*

The gravitational potential in three dimensions is given by

$$V_{\rm grav} = G_{\rm N}\,\frac{m_1 m_2}{r}, \tag{15.5}$$

where $G_{\rm N}$ is Newton's constant and r denotes the distance between the two masses m_1, m_2. The dimensions of various quantities are given by

$$\begin{aligned} [m_1] = [m_2] &= [M],\\ [r] &= [L], \\ [V_{\rm grav}] = [\text{Energy}] &= [M][L]^2[T]^{-2}, \end{aligned} \tag{15.6}$$

where $[M], [L], [T]$ represent arbitrary scales of mass, length and time, respectively. It follows, therefore, that

$$\begin{aligned} [V_{\rm grav}] &= [G_{\rm N}] \times \frac{[m_1][m_2]}{[r]}\\ \text{or}\quad [M][L]^2[T]^{-2} &= [G_{\rm N}] \times \frac{[M]^2}{[L]} = [G_{\rm N}] \times [M]^2[L]^{-1} \\ \text{or}\quad [G_{\rm N}] &= [M]^{-1}[L]^3[T]^{-2}. \end{aligned} \tag{15.7}$$

This determines the dimensionality of Newton's constant.

The dimensions of several other fundamental constants are:

$$[c] = [L][T]^{-1}, \quad [\hbar c] = [M][L]^3[T]^{-2}. \tag{15.8}$$

It therefore follows that

$$\left[\frac{\hbar c}{G_{\rm N}}\right] = \frac{[M][L]^3[T]^{-2}}{[M]^{-1}[L]^3[T]^{-2}} = [M]^2. \tag{15.9}$$

We can consequently associate a natural mass scale with the gravitational interaction:

$$M_{\mathrm{P}} = \left(\frac{\hbar c}{G_{\mathrm{N}}}\right)^{\frac{1}{2}} \approx \left(\frac{\hbar c}{6.71 \times 10^{-39}\,(\hbar c)\,(\mathrm{GeV}/c^2)^{-2}}\right)^{\frac{1}{2}}$$
$$\approx (1.5 \times 10^{38}(\mathrm{GeV}/c^2)^2)^{\frac{1}{2}} \approx 1.22 \times 10^{19}\,\mathrm{GeV}/c^2, \qquad (15.10)$$

where we have used the value of G_{N} from the Table of constants in Appendix E of the text. M_{P} is known as the Planck mass, and it defines an energy scale where gravitational interactions become important in particle interactions.

Similarly, we note that

$$\left[\frac{G_{\mathrm{N}}\hbar c}{c^4}\right] = \frac{[L]^6[T]^{-4}}{[L]^4[T]^{-4}} = [L]^2. \qquad (15.11)$$

This shows that there is a natural length scale associated with gravitational interactions, which is known as the Planck length:

$$\ell_{\mathrm{P}} = \left(\frac{G_{\mathrm{N}}\hbar c}{c^4}\right)^{\frac{1}{2}}$$
$$\approx \left(\frac{6.71 \times 10^{-39}\,(\hbar c)(\mathrm{GeV}/c^2)^{-2}\,(\hbar c)}{c^4}\right)^{\frac{1}{2}}$$
$$\approx (6.71 \times 10^{-39}(\mathrm{GeV})^{-2}(2 \times 10^{-1}\,\mathrm{GeV} - \mathrm{F})^2)^{\frac{1}{2}}$$
$$\approx (6.71 \times 10^{-39} \times 4 \times 10^{-28}\,\mathrm{cm}^2)^{\frac{1}{2}}$$
$$\approx 1.62 \times 10^{-33}\,\mathrm{cm}. \qquad (15.12)$$

The Planck length can also be obtained from the uncertainty principle, as follows. We note that the Planck mass also defines a characteristic momentum scale

$$p = M_{\mathrm{P}}c = \left(\frac{\hbar c}{G_{\mathrm{N}}}\right)^{\frac{1}{2}} c = \left(\frac{\hbar c^3}{G_{\mathrm{N}}}\right)^{\frac{1}{2}}, \qquad (15.13)$$

where we have used the definition of Planck mass from Eq. (15.10). This momentum can lead to an uncertainty in the position of the form

$$\Delta r \approx \frac{\hbar}{p} = \frac{\hbar}{\left(\frac{\hbar c^3}{G_{\rm N}}\right)^{\frac{1}{2}}}.$$

$$= \left(\frac{\hbar G_{\rm N}}{c^3}\right)^{\frac{1}{2}} = \left(\frac{G_{\rm N}\hbar c}{c^4}\right)^{\frac{1}{2}}$$

$$= \ell_{\rm P}, \tag{15.14}$$

where in the last step we used our definition of the Planck length.

Problem 15.3 *Ignoring, for the moment powers of $\hbar c$, Newton's law for n extra dimensions can be written as*

$$V_{\rm grav}(r) \propto \frac{1}{M_S^{n+2}} \frac{m_1 m_2}{r^{n+1}},$$

where m_1 and m_2 are the interacting masses, and M_S corresponds to the effective Planck scale for $n+3$ spatial dimensions. Assuming that these n extra dimensions are compactified over equal radii R, then $V(r)$ for $r \gg R$, that is, from the perspective of our three-dimensional space, becomes

$$V_{\rm grav}(r) \to \frac{1}{M_S^{n+2}} \frac{m_1 m_2}{R^n r}.$$

Now, using the fact that $M_S^{n+2} R^n$ must equal $M_{\rm P}^2$, calculate R in meters for $n = 1, 2, 3$, and ∞, with M_S set to the desired value of $\approx 1\,{\rm TeV}/c^2$. From what you know of Newton's law, is it possible to have $n = 1$? (Hint: Clearly, you cannot ignore $\hbar c$ in calculating R! Using the fact that $(Mc) \times (R) \approx \hbar$, and Problem 15.2, should enable you to get the right answers.)

In $(n+3)$ spatial dimensions, the gravitational potential has the form (this arises from the requirement that, for central forces, such as gravitation and the electric force, the field flux should be a constant)

$$V_{\rm grav}(r) = \alpha\, \frac{m_1 m_2}{r^{n+1}}, \tag{15.15}$$

where r is the distance between the two charges in the $(n+3)$-dimensional space, and α denotes a proportionality constant — much like Newton's constant in 3 dimensions. The dimensions of α can be determined from an analysis parallel to the one done in the last problem, and we obtain

$$\begin{aligned}
[V_{\text{grav}}] &= [\alpha] \times \frac{[m_1][m_2]}{[r]^{n+1}} \\
\text{or} \quad [M][L]^2[T]^{-2} &= [\alpha] \times [M]^2[L]^{-(n+1)} \\
\text{and} \quad [\alpha] &= [M]^{-1}[L]^{n+3}[T]^{-2}.
\end{aligned} \tag{15.16}$$

Recalling the dimension of $(\hbar c)$, we can write

$$[\alpha] = [M]^{-(n+2)} \times \left[\left(\frac{\hbar c}{c^2}\right)^{n+1} \times c^2\right] = \left[\frac{\hbar c}{M^2}\right] \times \left[\left(\frac{\hbar c}{Mc^2}\right)^n\right] \times [c^2]. \tag{15.17}$$

This suggests that we can introduce a characteristic mass scale M_S (similar to the Planck mass in 3 dimensions), and write

$$\alpha = \frac{\hbar c}{M_S^2} \times \left(\frac{\hbar c}{M_S c^2}\right)^n \times c^2. \tag{15.18}$$

When n of the space-like dimensions are compactified on a radius $R \ll r_3$, where r_3 denotes the large three-dimensional distances of space, the effective gravitational potential can be written (in 3 dimensions) as

$$V_{\text{grav}}^{(\text{eff})}(r) = \alpha \, \frac{m_1 m_2}{R^n r}, \tag{15.19}$$

where for simplicity we have identified r_3 with r. Comparing with the original expression for three dimensions, we can write

$$\begin{aligned}
G_{\text{N}} &= \frac{\alpha}{R^n} \\
\text{or} \quad \frac{\hbar c}{M_{\text{P}}^2} &= \frac{(\hbar c)^{n+1}}{M_S^{n+2} c^{2n}} \times \frac{1}{R^n} \\
\text{or} \quad R^n &= \left(\frac{\hbar c}{M_S c^2}\right)^n \times \left(\frac{M_{\text{P}}}{M_S}\right)^2 \\
\text{and} \quad R &= \frac{\hbar c}{M_S c^2} \left(\frac{M_{\text{P}}}{M_S}\right)^{\frac{2}{n}},
\end{aligned} \tag{15.20}$$

where we have used our previous definitions of the parameters. Furthermore, taking the scales

$$M_S = 1\,\text{TeV} = 10^3\,\text{GeV} = 10^6\,\text{MeV}, \quad M_{\text{P}} \approx 1.22 \times 10^{19}\,\text{GeV}/c^2, \tag{15.21}$$

we obtain for the radius of compactification in n dimensions:

$$\begin{aligned} R^{(n)} &\approx \frac{200\ \text{MeV} - \text{F}}{10^6\ \text{MeV}} \times \left(\frac{1.22 \times 10^{19}\ \text{GeV}/c^2}{10^3\ \text{GeV}/c^2}\right)^{\frac{2}{n}} \\ &\approx 2 \times 10^{-17}\ \text{cm} \times \left(1.22 \times 10^{16}\right)^{\frac{2}{n}}. \end{aligned} \tag{15.22}$$

We can now make a table for the values of R for different values of n.

n	R(cm)	Comments
1	3×10^{15}	ruled out by astronomical observations.
2	0.244	excluded by experiment.
3	1.06×10^{-6}	excluded by particle experiments
$\vdots$		
∞	2×10^{-17}